I͞A

Industrial Archaeology

A SERIES EDITED BY
L. T. C. ROLT

2
Iron and Steel

BY THE SAME AUTHOR

The Black Country Iron Industry:
a Technical History

The British Iron and Steel Industry:
a Technical History

Iron and Steel

W. K. V. Gale

Longmans

LONGMANS, GREEN AND CO LTD
London and Harlow
*Associated companies, branches and representatives
throughout the world*

© *W. K. V. Gale 1969*
First published 1969

SBN: 582 12649 5

*Printed in Great Britain
by W & J Mackay & Co Ltd, Chatham*

Contents

Author's Note		*page*	ix
Preface			xi
Note on Sources			xv

Chapter

ONE	The Nature, Properties and Forms of Iron	1
TWO	Prehistoric and Early Ironmaking up to 1700	10
THREE	New Processes and the Spread of Mechanization, 1700–1800	26
FOUR	Wrought Iron Ascendant, 1800–1860	40
FIVE	Wrought Iron Supreme, *c.* 1860	55
SIX	The Challenge of Steel, 1860–1890	69
SEVEN	New Products and the Impact of Electric Power, 1890–1945	84
EIGHT	Steel from 1945 to the Present	97
NINE	Ironfounders and Steelfounders	112
TEN	Smiths and Ironworkers	124
Gazetteer		139
A Select Bibliography		143
Index		147

List of Illustrations

facing page

1. Part of a wrought iron ball as withdrawn from the puddling furnace. (Author's collection.) — 14
2. Wrought iron bar partly cut and bent back cold, showing fibrous structure. (Author's collection.) — 14
3. Cross-sections of eighteenth century blast furnace. *Upper*, through the forepart; *lower*, through the tuyère. (Diderot et D'Alembert, *Encylopédie ou Dictionnaire Raisonné des Sciences, des Arts, et des Métiers*, 1751–1765.) — 15
4. Slitting mill, *right*; and flat rolls, *left*. (Diderot et D'Alembert, op cit.) — 15
5. John Wilkinson's Bradley works, Bilston, Staffs. (From a painting by Robert Noyes, 1836. By permission of the William Salt Library, Stafford.) — 30
6. Brymbo No. I furnace. An eighteenth century masonry blast furnace as it was at the end of the nineteenth century. (By permission of The British Iron and Steel Research Association.) — 30
7. Three nineteenth century blast furnaces charged from bank at rear, at Blist's Hill, Shropshire. (Author's collection.) — 31
8. The nineteenth century blast furnaces at Priorfields in Staffordshire, charged by hoist in tower on the left. (Author's collection.) — 31
9. Four nineteenth century blast furnaces at Cyfarthfa in Wales, charged by an incline between the middle two. (Author's collection.) — 46
10. Two nineteenth century blast furnaces at Netherton, Dudley, Worcestershire. One, on the left, is a normal masonry furnace; the other is of the iron-cased type. (Author's collection.) — 46
11. A typical pig bed at casting. (Author's collection.) — 47
12. The Earl of Dudley's Round Oak works at Brierley Hill, Staffordshire, a characteristic wrought ironworks of the mid-19th century. (Author's collection.) — 47
13. Water-driven tilt hammer at Wortley Ironworks, near Sheffield. (Author's collection.) — 62
14. Taking the ball out of a puddling furnace. (Author's collection.) — 62
15. Steam-engine-driven helve hammer used for shingling. Formerly at Stourbridge Old Forge, Worcestershire. (Author's collection.) — 63

LIST OF ILLUSTRATIONS

16 Shingling under a steam hammer. (Author's collection.)   63
17 A two-high merchant rolling mill of the late eighteenth century. From Wortley Low Forge, near Sheffield, now preserved at Wortley Ironworks by the Sheffield Trades Historical Society. (Author's collection.)   78
18 Hand rolling of sheet. (By permission of Richard Thomas & Baldwins Ltd.)   78
19 Part of an original Bedson continuous rolling mill, showing the alternate arrangement of rolls. (By permission of Johnsons (Mill St) Ltd.)   79
20 The last cementation furnace to work in Sheffield (closed 1952) at the Sheffield works of Daniel Doncaster & Sons Ltd. (By permission of The British Iron and Steel Research Association.)   79
21 Four modern blast furnaces, the 'Iron Queens': 'Queen Mary', 'Queen Bess', 'Queen Anne' and 'Queen Victoria' at the Scunthorpe works of the Appleby-Frodingham Steel Company. (By permission of The United Steel Companies Ltd.)   94
22 A 110-ton electric arc furnace at the works of Steel, Peech and Tozer, Rotherham. (By permission of The United Steel Companies Ltd.)   94
23 A 25-ton Bessemer converter, blowing, at the Workington Iron and Steel Company, Cumberland. (By permission of the United Steel Companies Ltd.)   95
24 Pouring molten steel at a continuous casting plant, Shelton, Stoke-on-Trent. (By permission of Shelton Iron & Steel Ltd.)   95
25 Discharge section of continuous casting plant, Shelton, Stoke-on-Trent. The cast billets emerge from the curved guides in a horizontal position. (By permission of Shelton Iron & Steel Ltd.)   110
26 Outgoing side of roughing stand of universal beam mill at the Appleby-Frodingham Steelworks, Scunthorpe. (By permission of The United Steel Companies Ltd.)   110
27 Outgoing side of finishing stand of universal beam mill at the Appleby-Frodingham Steelworks, Scunthorpe. (By permission of The United Steel Companies Ltd.)   111
28 Making a heavy shackle by hand forging. (Author's collection.)   122
29 Forging a crankshaft under the steam hammer. (By permission of Walter Somers Ltd.)   122
30 A modern 3,000-ton hydraulic forging press, with preset control of stroke, at Walter Somers Ltd., Haywood Forge, Halesowen, Birmingham. (By permission of Walter Somers Ltd.)   123
31 Some of the many shapes in which iron could be rolled, in addition to the ordinary bars and sections. These samples of fancy iron are in the author's collection.   123

Author's Note

No work of the present type could ever be the work of a single individual. I acknowledge gratefully the help I have received over the years from managements and men, in all parts of the country, who gave freely of their time and allowed me to draw on their experience and skill to help me to understand their often complicated but always fascinating art. They are far too numerous for me to name individually (and indeed many of their names were unknown to me). It would be invidious to single out a few for mention, but if my benefactors must, perforce, remain anonymous, at least my debt shall be recorded.

I am grateful, too, to those firms and organizations which allowed me to use their illustrations and last, but far from least, to my wife, who typed the whole of the manuscript, and a great deal more as well.

W.K.V.G.

Preface

Iron and steel are universal. They are now so ubiquitous that they are generally taken for granted. Yet the present industrial civilization of the world is firmly founded on iron in its many forms, of which steel is today the most important. Iron has been with us a long time, is now firmly ensconced, and looks like staying for many years to come, though this does not mean that it is unchanged. Far from it; many forms of iron have evolved over the years and some are still being developed. Some have become extinct, like blister steel, or virtually so, like wrought iron. Others, like steel, have grown in importance and have themselves evolved into large groups of materials, with many different characteristics and little in common but their natural origin.

It would be unrealistic to say that man could not do without iron. He did so once and he may eventually have to do so again, for the world's resources of ironmaking materials, though vast, are not inexhaustible. Indeed, some of the best are already in short supply in some parts of the world, though this is no cause for concern, for techniques have been developed for dealing with others hitherto unusable. In the long term, too, more use will probably be made of substitute materials and it could be that man will have found out how to exist without iron long before raw material shortages bring compulsion. But all that is very much in the future, and it would be extremely difficult to manage without iron today. It is employed in the making of all the machines we use for transport and manufacture. So much is obvious. What is not, perhaps, quite so obvious is its value to the modern builder and civil engineer. We are so often told, and visually reminded, that concrete is the building material of today that we are inclined to forget the thousands of tons of steel reinforcement that lie hidden in our tall modern buildings. If all the sources of iron were to be cut off suddenly now, the world would be in a sorry state.

It is a measure of the importance of iron, the most widely used of all the metals, that Great Britain alone produces more than 26 million tons of it (largely in the form of steel) a year, and this country is far from being the world's largest producer of iron, though it was once. The annual world production of iron and steel is something like 425 million tons, a sizeable output by any standards and one which, given stable world conditions, is likely to increase.

Because of its worldwide importance, iron has attracted the attention of many writers but while the present and future have often been considered from a technical point of view, the history of the subject has in the main proved of more interest to economists. Economic and sociological history are important (though not all-important as is sometimes asserted) but it must not be forgotten that products, plant and processes are fundamental to any manufacturing industry.

The object of the present book is to look at iron from the point of view of the people who made, and make, it and to examine in some detail the techniques, machines and products which have made iron and steel what they are today. To understand properly what iron is, where it comes from, how it has been developed and why, from the earliest times to the present day, inevitably involves consideration of technical matters and that is the purpose of this book. Because the book is primarily concerned with plant and techniques, statistics of production, which are of more concern to the economic historian, find no place in it. Figures are given only when they are needed to indicate or emphasize a technical development.

Iron can be defined quite simply, and such a definition will be the basis of the first chapter, but that is all that is simple about the subject, for iron is not one thing but many. It can be so prepared that it appears in numerous guises, capable of an almost endless variety of tasks. One important point must be made. The products of the iron and steelmaker are of no practical value in their own right. They are simply the raw materials of other industries. Such things as pig iron, castings, forgings, rolled bars, sections, sheets and plates, must be worked on by other manufacturers before they have a useful function. How this working-up is done is not within the scope of this book, which is solely concerned with the *making* of iron and steel. There is, however, a certain amount of overlap. Neither castings nor forgings, nor certain other products such as nails, chains and tubes are made by the iron and steelmaker in the proper sense of the words. But they are all made by plant and processes similar to those used in iron and steel production, and some of them were, at one time, actually made in the ironworks. Moreover, most of them fall readily into our category of raw materials for other industries. They are therefore dealt with in a final chapter.

Note on Sources

The main dates and facts outlining the history of iron and steel during what may be called the industrial revolution period (with which this book is concerned), are readily ascertainable from various easily accessible sources. Similarly, the principal social and economic consequences of the development of iron and steel from an unimportant craft to a great industry are related in many published works. I have listed in the bibliography a selection of books which will not only confirm this information but will enlarge upon it and indicate further sources. I have therefore felt it unnecessary to give source references for my main dates and facts.

Dates and facts alone, however, are of little use, unless one is content to have just a cursory knowledge of the subject. From the earliest days of my own studies I found it a source of irritation that while I could often discover quite easily what happened in a particular field and when it happened, there appeared to be no answer to the question of why it happened. Thus it was easy enough to find out that the Bessemer process, startlingly successful at first, soon fell into disrepute, from which it did not recover either quickly or easily. Further enquiry elicited the fact that the iron used by some of Bessemer's licensees did not suit the process as it then was. But why, fundamentally, did the failure occur? After all, the Bessemer process did eventually become a tremendous factor in the industry. Again, although the joist, a section very useful to structural engineers, was known in Britain from about 1850 onwards, in the heyday of wrought iron very few joists were rolled in that material. Was it unsuitable for joists and if so, why? In fact, of course, wrought iron as a material would have been perfectly satisfactory as joists; the reasons why such sections were not made in quantity were related purely to manufacturing techniques. There were many more of these puzzles (and some, it may be added, still remain), but one more example will suffice here. Why did mild steel supersede wrought iron? Was it a better material? At first it was not, and the facile answer supplied by most of the textbooks was that it was cheaper to make. So it was, but that is far from being the whole story. It is this sort of puzzle which I have tried to deal with in the book.

To try to answer some of these questions I read every published source I could find, only to discover that I could not understand many of them. Contemporary accounts of iron and steel in Britain prior to the eighteenth century are negligible, but mention should be made of *Metallum Martis*, Dud Dudley, 1665, reprinted in facsimile in 1854, if only to dismiss it as a curio. It contains nothing of technical value. Neither do *A Treatise of Metallica*, S. Sturtevant, 1612, or another work of the same name but by J. Rovenzon, 1613, both of which were reprinted and bound with *Metallum Martis* in 1854. Textbooks were rare in the eighteenth century and most of those there were are very hard to find. Such a book is Réaumur's *Memoirs on Iron and Steel*, published in French in 1722 and only translated into English and published (in America) in 1956. The iron and steel section (*Forges ou Art du Fer*) of the famous *Encyclopédie* of Diderot and D'Alembert, published between 1751 and 1756 has never been translated into English and is not easy to obtain, and the only public library listing a complete copy of *Voyage Métallurgique en Angleterre*, Dufrenoy et al, 1837, was the British Museum library. My own study of this work was done on a volume borrowed from Paris. It might be added here, in parenthesis, that some of the best accounts of the British iron industry in the eighteenth century are French.

Textbooks of the nineteenth century are more numerous but not always any easier to obtain. *Metallurgy of Iron*, J. Percy, 1864, is to be found in many libraries and sometimes comes on the second-hand market; this great classic of its time is a mine of information, but needs a fair technical knowledge to appreciate. *Papers on Iron and Steel*, D. Mushet, 1840, is more difficult to find. On the other hand, *Guide to the Iron Trade*, S. Griffiths, 1873, while fairly easy to discover in libraries, is wholly unreliable in parts. A few books, even when found, prove to be unintelligible unless the reader has specialized knowledge of the period covered. Such a book is *Guide to the Iron Trade*, B. Legge, 1840, and in any case it is extremely rare.

Then there are records, manuscripts, maps and plans (and a few rare books) which are not accessible for study, because they are in private hands. I have been privileged to see and use many such records, and the position is improving for students generally as several collections are now in the care of public libraries and county record offices. Institutions of this kind are always worth approaching; often they have collections which are virtually virgin material as far as the historian is concerned. Glamorgan County Record Office, for example, has a great and valuable collection of papers from Dowlais Ironworks, some of which have been published by the Council in *Iron in the Making*, 1960.

There were many papers, too, which have been wantonly destroyed even in recent years and although I have in some cases managed to get a few notes from them, it would be pointless to list them as they no longer exist.

There are still many sources to be found by the reader who will search diligently, and there is no doubt that some as yet unknown will turn up as private owners decide to hand them over to public libraries and record offices. But a word of warning is needed. I said earlier that I found some of the records unintelligible, and there was only one answer to this problem. My failure to understand a technical book was generally my own fault; not always, though, sometimes the author himself was wrong or obscure. However, a first requisite for a full understanding of technical records was, and still is, a thorough knowledge of the basic principles of iron and steel making. This I have had to acquire in the course of my work, and since the basic principles of iron and steel making depend on fundamental natural laws, which have always applied, even modern technical knowledge can be projected backwards to old processes with valuable results.

This brings me to my final, and in some ways, my most important source of information. For years I have lived close to the industry and have never missed the opportunity to handle the tools on the job, to puddle, to roll, to forge, to mould and at least to assist in any way possible in the making and shaping of iron and steel. Many processes now obsolete survived into my time, and I learnt them by doing them. I took the opportunity, too, to talk to old operatives in the trade, who told me of jobs that had gone before I was born, and, by collating and analysing many such recollections (which were sometimes contradictory!) I was able to get a clear and reliable record of what went on. To some extent, therefore, much of the material in this book is a kind of distillation of the personal experiences of many people. They were people who actually did the jobs, not those who compiled figures about them, and records obtained in this way, if examined from a strictly technical point of view by a suspicious person (which a historian ought to be) can be accepted as reliable. Of course, personal recollections of the earlier periods are more difficult to obtain, but diaries and notebooks can give a lot of useful information if studied critically.

Today scientific method rules more and more, in some works absolutely. In the past it was not always so and even the textbooks of the period can mislead. As a man of great practical and theoretical renown once said, 'The trouble is, nobody has yet taught the furnaces to read the books.' So a lot of what I have said in this book has come from practical experience (my own and that of other people) actually on the job. There is still scope for this sort of approach. Although some of the processes I have described have now disappeared, others remain, and they too will disappear in time. Some of them are likely to go soon and suddenly. They can be recorded by those who will take the trouble, but it is hard work and there is no royal road to success. If it is hard, though, it is always rewarding and sometimes exciting.

CHAPTER ONE

The Nature, Properties and Forms of Iron

Iron is an element, with the chemical symbol Fe, an atomic weight of 55·85, a specific gravity of 7·87 at 20°C, and a melting point of 1,535°C. Many other physical facts are on record, but the above will suffice for our purpose. What we are more concerned with is what can be done with iron to make it useful to man, for although it is the fourth most widely distributed of all the elements, in the form in which it is found in nature it is of no use at all. The source of iron, iron ore, is quite simply a rock or stone, which has little or no value in that form. It is true that a usable form of iron, which could be hammered out into such things as tools, has been found in various parts of the world, but this, meteoric iron, which is really a natural alloy of iron and nickel, is so rare as to be of no commercial value. Samples of meteoric iron can be found in museums, and perhaps in the very distant past a few specimens were worked up into some useful form. But for all practical purposes meteoric iron can be regarded only as a curiosity. The iron of commerce comes from one or other of the naturally occurring iron ores, which are found in various parts of the world, and differ considerably in their degree of purity.

Iron, as we have said, is very widely distributed, but only in certain forms is it of commercial value. In many places the amount of metallic iron which could be extracted from the earth is far too small to be economic. Iron betrays its presence in many places by the colouring it gives to the soil, but unless the quantity of metallic iron which can be recovered is more than 15 per cent, at the very lowest, it is not worth working. Sometimes, too, there will be other elements present to such an extent as to render the mineral unworkable and there are known deposits which have hitherto been unusable because of their physical condition. Such are the iron-bearing sands of New Zealand, which could not be treated by any known method until recently, when means have been devised to deal with them.

Many elements have a particular affinity for other elements, and iron shares this feature. It unites very easily with oxygen, the most abundant of all the elements, as the familiar rust indicates, for rust is only an oxide of iron. Left to itself, unprotected, iron will always turn into an oxide, faster or slower according as conditions are more or less favourable. All the iron ores of commerce are basically oxides of iron, though as found in the earth they are combined chemically with other elements as well, in varying proportions and mixed with earthy waste or gangue. It is the business of the ironmaker to remove the waste and unwanted elements to the degree required to produce the form of iron he needs. He cannot remove all traces of other elements and would not if he could, for some of those elements confer useful properties on the iron, as will be seen. As a matter of interest, pure iron, that is iron chemically free from all other elements, would be extremely difficult to produce and would have no commercial value anyway. What the ironmaker has to do is to control the other elements in the way he wants, to get the desired result. This involves not only the removal of some elements, at some stages of the process, but sometimes putting them back at other stages; this may seem odd, but it is often the easiest way to get the exact percentage required.

There are three basic kinds of iron ore. The best, from the point of view of iron content, are the magnetite ores, which contain up to 65 per cent or even more metallic iron. Unfortunately, they are also the rarest. The next best, for iron content, are the hematite ores, which have 50 to 60 per cent iron. They are widely distributed throughout the world, including Britain, but the quantities available in this country are small. Then there are several different kinds of ore which can be grouped together as low grade. Some of them are very poor in iron content, ranging as low as 15 per cent, when they are known as lean ores. These include some called carbonates, which are not simple oxides, but although they have to undergo special treatment they are nevertheless dealt with finally as oxides, and the general proposition we have stated—that the ores of commerce are iron oxides—remains sufficiently accurate for our purpose.

Given an iron ore, which is primarily an oxide mixed physically

THE NATURE, PROPERTIES AND FORMS OF IRON

with some earthy materials and chemically with other elements, how does the ironmaker set about turning it into metallic iron? He has to remove the oxygen, a process known chemically as reduction and from a practical point of view, today, as smelting. Iron and oxygen have, as we have said, a strong affinity for each other, but luckily there is another element, and a common one, which has an even greater affinity for oxygen. This is carbon. If iron ore is treated suitably in contact with carbon, the carbon unites with the oxygen and goes off in the form of a gas, leaving the iron behind. That is iron smelting reduced to its simplest terms. Smelting is of course much more complicated in practice and some of the processes involved are very complex indeed, but the fundamental facts remain; iron ore (oxide) is reduced by reaction with carbon to a metal and a gas. Stated in elementary chemical terms the reactions in smelting a hematite ore are:

$Fe_2O_3 + 3C = 2Fe + 3CO$; that is; iron ore + carbon = metallic iron + carbon monoxide gas.

In practice the reaction always takes place hot. It is faster this way, and the commercially available forms of carbon (coke today, charcoal formerly), burn readily to generate heat. So the reactive element, carbon, is also the source of heat which makes the process fast enough to be a commercial proposition. Because the process is carried out at great heat, the metallic iron always emerges very hot. Today it is in fact molten. Formerly this was not so, as will be seen. In the ancient processes the iron was hot but not molten, and its physical characteristics were different. But today, and for the last few hundred years, the first product of the ironworks is molten iron.

Such, in essence, is the production of iron from the ore, but it is not the end of ironmaking; it is only the beginning.

There are three basic commercial categories of iron, wrought iron, cast iron and steel, wrought iron being by far the oldest and steel the newest. Wrought iron is now effectively extinct (the so-called 'wrought iron' gates and decorative domestic items of today are not made of wrought iron at all, but of mild steel). But it was for thousands of years the only form of iron used by man and for hundreds of years, even when other types of iron were in use, by far the most important. It will play a large part in the present volume.

Wrought iron is the commercially pure form of iron, containing as little as possible of other elements. Two of these elements in particular, are highly undesirable in wrought iron (and, for that matter in steel, where they can also appear and cause trouble). These elements are sulphur and phosphorus. Sulphur, even in very small percentages, causes wrought iron to be red short, or hot short, that is it is incapable of being worked to shape while it is hot, crumbling or breaking instead of taking the form the worker is trying to give it. Phosphorus, on the other hand, causes the iron to be cold short, or weak and brittle when cold. Either condition renders wrought iron of little use for most purposes and quite useless for some. Red shortness, for example, would make it impossible to form the links of a wrought iron chain, while cold shortness would render the chain liable to break and therefore dangerous in use. Both sulphur and phosphorus are present in some iron ores, and can be picked up during the manufacture of iron in more than one way. Their presence in the finished product has to be guarded against with great care.

Wrought iron has a fibrous structure. If a piece of wrought iron bar is nicked on one side with a chisel and bent back on itself the fibres of the metal will tear and the bend will open out; it has the appearance then of a piece of wood, the fibres lying along the length of the bar. In practice, the distribution of the fibres depended on the way the iron was made and worked, and, as will be seen, the methods of manufacture were varied to some extent according to the use to which the iron was to be put. This, however, is looking ahead.

For the present it is sufficient to remember that wrought iron has a fibrous structure. It is ductile, that is it can easily be worked into shape by hammering or rolling while it is hot. Further working, such as drawing into wire, can be done while the iron is cold, and it can in fact be hammered to shape or rolled while it is cold, though more force will be needed and cold working causes the iron to become crystalline and lose a great deal of its strength. Cold working other than wire drawing, is not done in practice, except for a certain amount of cold rolling, and this is followed by heat treatment to restore the metal to its original condition. The working of wrought iron into the required finished shapes was therefore almost always done hot (except in the case of wire).

A further, and very useful characteristic of wrought iron is its ease of welding. Heated to the correct temperature, about 1,300°C which shows as a white heat, a piece of wrought iron can be united firmly or fire welded to another piece at the same temperature by hammering or squeezing. This useful property was exploited to the full in both the manufacture and the manipulation of wrought iron and for some years gave it an advantage over its rival steel, which did not weld so easily or so successfully. Today, of course, modern welding processes have set at naught such advantages as wrought iron had, and steel (and many other materials, ferrous and non-ferrous) are welded to very exacting specifications.

The welding properties of wrought iron are assisted by the presence in the metal of threads of slag mixed physically (not chemically) with the fibres. This slag, or cinder, which is picked up during the manufacturing process, is technically an impurity, but its presence is the reverse of harmful. It helps in welding because it melts at a temperature lower than that needed for the welding operation, and protects the iron against oxidation, which would cause bad welding. As the faces of the iron are squeezed or hammered together, the molten slag is pushed out, and the clean iron faces are united. The slag is also considered to help in inhibiting corrosion, as the non-metallic threads, scattered throughout the metal, prevent corrosion from striking deeply below the surface, where it starts. Corrosion resistance was another very valuable characteristic of wrought iron which was at one time used in its favour *vis-à-vis* steel. Again, this has ceased to be of significance today, when various types of ferrous metals have been devised which resist corrosion very successfully.

To sum up wrought iron, it is the commercially pure form of iron, fibrous in structure, easily forged and shaped while hot, easily welded, and with good corrosion resistance. It is fairly strong in tension (that is it resists forces tending to pull it apart) but not so good in compression (when the forces are tending to squeeze or compress it). A good wrought iron has a tensile strength of up to 20 to 35 tons/in^2. Although it can be worked while hot it does not melt easily. To melt it requires a temperature of between 1,500°C and 1,600°C, which is about the same as that of steel; the melting points of both metals vary according to their composition, but these figures are useful approximations. In the wrought iron age such temperatures were difficult to achieve and in fact they were unnecessary, since wrought iron was never melted. It was always worked by forging or rolling, at temperatures well below the melting point.

Cast iron (also referred to as pig iron or blast furnace metal), the next oldest to wrought iron, was, and since it is still made and used extensively, is, very different from the oldest form of iron. It can be defined simply as an alloy of iron and other elements, chiefly carbon, phosphorus, silicon and manganese, these can be present to a total of about 10 per cent with carbon as the primary one. Cast iron is a crystalline metal, easily melted and capable of being cast molten into moulds, the shape of which it will take and retain when cold. It melts, in fact, at 1,140°C and it is of interest to note that the addition of carbon lowers the melting point of iron and vice versa. This fact will be seen, later, to have a bearing on the production of wrought iron. Cast iron is poor in tension, having a tensile strength of no more than 5 to 8 tons/in^2, but very good indeed in compression, being able to withstand very heavy crushing loads. It cannot be forged, rolled or shaped in any way either hot or cold (other than by casting it into a mould), and it cannot be welded in the same way as wrought iron. Today cast iron can be welded by modern electric and gas-welding methods, but even now welding it is not a simple matter, and requires special techniques.

Cast iron is now, and has been for some considerable time the first product of the iron and steel making processes. Some of it is used for making castings, but the bulk goes for further processing, today into steel, formerly into wrought iron. Briefly, then, cast iron is an iron alloy containing carbon up to 3 or 4 per cent or a little more, with small percentages of other elements. It is crystalline, brittle, poor in tension and good in compression, and can be cast but not wrought. One final fact about cast iron must be noted. It can now, by special treatment, particularly by some techniques evolved in recent years, be made to acquire one or more of several special properties not otherwise present in normal cast iron, such as ductility. But such special treatments produce high-duty cast irons, which will be dealt with in the appropriate place. For the present, cast iron should be considered as defined.

Steel is much more difficult to define; it is not one metal but many, each with widely differing characteristics. Fundamentally, steel is an alloy of iron and carbon, but so is cast iron, which is very different, and this simple statement, though true, is not sufficient. At one time steel was considered to be any form of iron which could be hardened, but this, again, is not good enough; it is not, nowadays, really true, either. For the purposes of definition steel can be divided into three basic types, mild steel, carbon steel and alloy steel.

Mild steel is the commonest of all, being used for countless types of engineering and constructional work where no special qualities such as great strength, heat resistance, hardness or corrosion resistance are needed. It is really, in the chemical sense, closely akin to wrought iron, for it is iron with the minimum quantity of other elements which can be achieved commercially. There are no slag threads, however, and mild steel, which is ductile and can be forged, rolled and worked just like wrought iron is not very good at corrosion resistance, and while it can be fire welded, it is inferior in this respect to wrought iron. Mild steel contains up to about 0·25 per cent carbon and often less, such lower-carbon qualities are usually called very mild steel. The tensile strength of mild steel is usually rather better than that of wrought iron.

Carbon steel can itself be subdivided into two classes, medium-carbon and high-carbon. The addition of a very moderate amount of carbon to steel has a very marked effect on its physical properties. It now becomes possible to harden it, by heating it to red heat and cooling it rapidly, usually by plunging it into cold water. In this condition alone, however, it is of little or no use, for it is not only hard, it is what is known as glass hard, and is so brittle that it could be broken quite easily. But if it is first hardened, then reheated to a lower temperature and cooled again, the steel undergoes a change known as tempering. The degree of hardness is reduced or drawn, and the steel, while retaining a degree of hardness, is much less brittle and much tougher. By varying the degree of reheat temperature and cooling at the appropriate point, so the degree of hardness can be varied between glass hard and almost, but not quite, soft.

Carbon steel with a carbon content of more than 0·25 per cent and up to 0·5 per cent is known as medium carbon steel. If the carbon content is above 0·5 per cent and up to the normal limit, about 1·4 per cent, the steel is known as high-carbon.

Carbon steel is older than mild steel and the knowledge that carbon could confer the property of hardening on iron is older than both. At least, it was known even in ancient times that it was possible to diffuse a certain material into the outer surface of wrought iron and so give a hard skin, though only a few thousandths of an inch thick.

This is known as case hardening, and it is still used widely, many steel components being case hardened for wear resistance.

Until about a century ago mild steel and carbon steel were the only types of steel used. Then came the first alloy steels, which will be considered in more detail in the correct place, chronologically, in our story. For the present it will suffice to say that by deliberately adding certain elements other than, or in addition to, carbon to steel very many special and valuable properties can be given to it. Thus, nickel and chromium confer on it the property of corrosion resistance, and manganese makes it resistant to abrasive wear.

By varying the chemical composition of steel, and by subjecting it to appropriate heat treatments to vary its metallurgical condition, it is possible to obtain a wide range of special properties besides hardness, corrosion resistance and toughness. It can be made easy to machine, to resist high temperatures and to resist fatigue (that is the repeated application and removal of loads). To do these things sometimes requires quite a high percentage of other elements, and as some of these are expensive, the resulting steels are also costly. Consequently, they are only used where they are absolutely essential. In some cases, for instance, in the aircraft and nuclear power industries, the alloy steels are often more alloy than steel; that is they have more alloying elements than iron, and they should not really be called alloy steels at all. Any alloy containing less than 50 per cent of iron is not strictly a steel. But these special alloys are made by steelmakers, using steelmaking plant and techniques, and they are usually listed with the alloy steels. In any case, these alloys, though important to engineers, are only needed in comparatively small quantities and do not really affect the issue.

Such is an outline of the basic forms in which iron and steel have appeared from time to time in history. Some have become extinct, some are new, some are still undergoing change. Still others are as yet in the development stage and are not of commercial interest at present. The industry has seen many changes, not only in its methods but in its products, and it is our intention to look at those changes. Like many other industries ironmaking developed slowly at first. There were long periods when little or nothing of importance happened at all. In more recent years change has been an increasingly important theme and it is still going on.

But for the present it is only necessary to remember that the fundamental forms of iron are wrought iron, cast iron and steel. Wrought iron is no longer of commercial interest, but both cast iron and steel are produced in vast quantities and in almost endless varieties, not only in their physical qualities but in their shapes, too. Castings, forgings, rolled and drawn products all play their part in modern life.

They are so multitudinous that it would be impossible to list even a fraction of them. Only a few typical ones can be mentioned. Modern cast iron can be seen in the motor-car cylinder block; in many hot-water radiators (though some are fabricated from steel sheet); in many machine parts such as lathe and other machine-tool beds, pillars, columns and gearboxes. Older iron castings can be found in hollow-ware and domestic grates; in bridges, railways, and structural members of buildings. Steel is to be seen everywhere as joists, channels, angles and tees in structural work; as sheet in the pressed and welded bodies of motor-cars, and in refrigerators and gas and electric cookers; as bars in machine parts and railings; as nails, screws, nuts, bolts and washers; and as tinplate in cans, containers, and decorative ware.

Carbon and alloy steels are on the whole not so obvious in an everyday world, except stainless steel, that is. Stainless steel is increasingly in evidence in cutlery and kitchen-ware, in table-ware and as a decorative finish. Large quantities are used in the food and drink industries. Alloy steels, too, though not usually seen, are found almost everywhere, from motor-car gearboxes to the tiniest wrist watch, from jet aircraft to a good quality penknife.

Wrought iron is hard to find nowadays. Many of the things now made of steel were formerly of wrought iron, but they have mostly disappeared. It can still be seen here and there, however, and the best examples are to be found in the fine decorative work, gates and railings now mostly to be seen in museums or in old houses. Wrought iron also survives in certain early railway bridges, notably the Britannia bridge over the Menai Strait, and the Royal Albert bridge over the river Tamar.

CHAPTER TWO

Prehistoric and Early Ironmaking up to 1700

Prehistoric ironmaking is a matter for the archaeologist rather than the technical historian, but for the sake of completeness something must be said about it briefly here. Neither the date nor the place of man's discovery of the art (for such it was) of ironmaking are known. Archaeologists using newly developed techniques have done much in recent years to elucidate some of the mysteries and their work is continuing. But for the present it is true to say that any account of the early days can be no more than speculative.

It is more than possible that man found out how to make iron by accident, and that the discovery was made independently at various times and places. The making of fire by artificial means was one of man's first technical achievements, and it is of interest to note that his fuel, wood, converts to charcoal, a very good form of carbon, in the fire. If a fire were built surrounded by stones of iron ore, which in the right areas it could easily be, conditions could readily arise in which the heat of the fire, fanned by the wind, would cause the charcoal to unite with the oxygen in the natural oxide ore and so deoxidize or reduce it, leaving behind a rough and ready form of wrought iron.

It would be obvious to an observer that some at least of the stones had changed their appearance, and sooner or later somebody would experiment with them. Striking two of the changed stones together, for instance, would disclose a very noticeable feature; the stones would give out a different sound and they would not shatter or chip. On the contrary, they would tend to bend or dent. If the experiment were tried while the stones were very hot the effect would be even more remarkable. They could be shaped, or forged, into, say, a knife or an axe-head, which, when cold, would retain its shape and be tough enough to withstand quite a lot of heavy work. A reasonable assumption is that somebody would be curious enough to collect a few likely-looking stones and heat them in a wood fire deliberately, to see what happened. The success of experiments of this type would naturally lead to further trials, to careful selection of those stones which experience showed gave the best results, and to attempts to heat them more effectively. So a technique emerged, which became the forerunner of iron and steelmaking today.

All this, as we have said, is purely speculative, though it is very reasonable to assume that something like it did in fact happen.

But if the origins of the process are a matter for speculation it is a fact that over 4,000 years ago man had found out how to effect direct reduction of iron ore to wrought iron, and that the metal was being made in various parts of the world. The process was simple, the equipment crude and the product of varying quality. Moreover, since the scale of production was tiny, the metal was scarce and expensive and its use was restricted. Still, the iron industry existed in embryo. From that time onwards ironmaking has been carried out continuously, though for centuries progress was very slow indeed.

For many hundreds of years there was virtually no change in production methods, though the art of casehardening was mastered, and the scale of production increased a little through the use of slightly bigger and better furnaces. It is worth recording, incidentally, that the same primitive methods which were practised before the Christian era began lingered on for centuries in some parts of the world and that there are a few places even now where these ancient methods are still in use.

It was not until the latter half of the fifteenth century that any radical change in techniques came about, and even then the change was gradual and confined to a few parts of western Europe. From then on, ironmaking (and later, steelmaking) became a two-stage process or, as it is known, in distinction from the older direct method, indirect reduction. Before dealing with this development, however, it is necessary to take a look at the ancient method, for the new one, radical though it was in many ways, was only a division of the old process into two distinct parts; the end product remained the same.

Direct reduction of iron from the ore is fundamentally very simple, though a fair amount of skill was needed to carry it out. There are many processes in ironmaking where the desired effect cannot be *seen* to be happening, and it is only in recent years that instruments and special techniques have been devised for scientific and often automatic control of a complete process. Even now things sometimes go wrong. In the past the successful control of a process was largely a matter of acquired individual skill. Often enough—indeed usually—the workers only knew *how*; they did not know *why*, they did certain things at certain times or in particular ways.

This was the case with direct reduction of iron from the ore, and misjudgement, or sometimes just misfortune, could result in the work of hours being wasted because the product was useless. It will be remembered that wrought iron, the desired product, is simply metallic iron with the absolute minimum of other elements, and that it can be produced by deoxidizing, or reducing, the oxide ore by heating it in contact with carbon. However, it is also true that if the heating is kept up for a sufficient period, and if the heat is raised to a high enough degree the iron, having got rid of its oxygen, will begin to take up some of the carbon, and will become cast iron. This, to the early workers and, in fact until means were found for first converting it to wrought iron in the fifteenth century and later for using it in its own right, was useless. It was brittle and quite unsuitable for tools and weapons. So the accidental production of cast iron which, in the right circumstances, was all too easy, was something to be guarded against with great care.

The furnaces (or bloomeries as they were called) used for direct reduction were all fundamentally the same, though they varied quite a lot in their actual physical shape and they grew in size a little as the years went by.

Essentially the direct-reduction furnace was an enclosed space in which charcoal was set alight and kept burning by means of bellows, and was fed with a little iron ore at a time by the man in charge. It could be simply a depression in the ground lined with some heat-resisting material such as clay and covered with a dome of the same material, a small hole being left in the top for the egress of gases (principally carbon monoxide from the chemical reaction of oxygen in the ore with the carbon of the charcoal) and for feeding in the iron ore as required.

Alternatively the furnace could be built above ground in the form of a short stubby chimney and made of clay, or of pieces of fire-resisting stone such as sandstone, bound together with clay. In either case, in addition to the gas exit/charging hole at the top there was a further aperture at the bottom, normally closed with a plug of ashes, through which the finished iron could be extracted at the end of the working cycle. Through this plug a baked clay nozzle or tuyere projected into the bottom or hearth of the furnace, and an air pipe or pipes brought the blast from the bellows into the tuyere. To take out the piece of finished iron the air pipe (often of wood but sometimes of clay) was removed, the tuyere was withdrawn, and the plug was broken out. Then, by scraping away the burning charcoal which surrounded it, the piece or bloom of iron was exposed and could be removed.

The bloom, which gave the name bloomery to the furnace, was of spongy texture, mixed with molten slag, and was at once beaten on a flat stone to expel surplus slag and to consolidate and weld together the iron. It was then ready for forging to the required shape, though it often needed reheating at least once or twice before the job was finished. Reheating was done in a simple charcoal fire, probably using the same bellows, the fire being not unlike the once common blacksmith's hearth, which in fact it was, though smaller.

Blooms made by direct reduction were small, only a few pounds of iron resulting from several hours' work. The process, too, was intermittent. When the workers judged that the bloom was ready, the forepart of the bloomery was broken out and the bloom was removed. Then the bloomery had to be cleaned out and any serious cracks or damage to the structure had to be repaired before the fire could be lighted again.

However, the equipment was simple to make, using materials dug out of the earth, except for the bellows. Because clays and heat-resisting stones vary in nature and occurrence in different places, the actual structure of the bloomeries varied, too, as we have said. It was a case of making the best use of what materials could be found locally. The same applied with the bellows. These, again, varied in type and shape according to circumstances. But necessarily they had to be simple.

A typical bellows consisted of a hollowed-out block of wood closed at the top by a piece of animal skin. This skin could be moved up and down, by hand or, very often, by simply treading on it, so compressing air in the bellows body, the air issuing from a small hole in the side into a pipe leading to the bloomery tuyere. A cord and a springy piece of wood pulled the bellows skin up again after it had been pressed in, allowing a fresh charge of air to enter through a hole in the skin; this hole was closed by the hand or foot of the worker on the pressure stroke. By using two or more bellows something approaching a continuous blast was kept up until the process was finished.

Such was the primitive bloomery. It worked but it had obvious limitations, not the least of which was its small scale. Because it was manually operated it could not be increased in size to any useful extent and it was only when the whole process changed and some rudimentary mechanization was applied that there was any significant rise in output.

This was brought about by the introduction of the indirect process in which the iron ore was first reduced to cast iron in a new type of furnace—the blast furnace—and then treated in a separate and distinct furnace—the finery—to remove the carbon (or to decarburize it), so producing the desired wrought iron. There is some obscurity in the early history of the blast furnace, and it is not known either who invented it or where, exactly, the first one was built. It is known, however, to have originated somewhere in the neighbourhood of Liège, in what is now Belgium, at some time before the year 1400, for it is recorded that one or two blast furnaces were in operation in the area in that year. No blast furnace was seen in Great Britain for some time after that, in fact it was not until about the year 1500 (or a year or two earlier) that the first one appeared in this country.

The blast furnace marked the first fundamental change in the long history of ironmaking. It was revolutionary in two respects: firstly it enabled much larger outputs to be achieved (several hundredweights in twenty-four hours instead of a few pounds); secondly it was continuous in operation. This second factor meant that some form of mechanization was necessary, and the requirement was met by means of water power. It also meant that a further element entered into the siting of the iron industry, for blast furnaces needed not only ore and charcoal but a stream or river to supply the power for blowing the bellows. The muscular efforts of men or animals could not serve; the cost would be too great. Neither could the windmill be used, since the process was continuous and the wind is not. Only water power would do and it remained in use until steam power displaced it. So the iron industry grew at first in such areas as the Weald in Sussex and Surrey, where nature had pro-

1 Part of a wrought iron ball as withdrawn from the puddling furnace

2 Wrought iron bar partly cut and bent back cold, showing fibrous structure

3 Cross-sections of an 18th century blast furnace. *Upper:* through the forepart, *lower:* through the tuyere

4 Slitting mill *right*, and flat rolls *left*

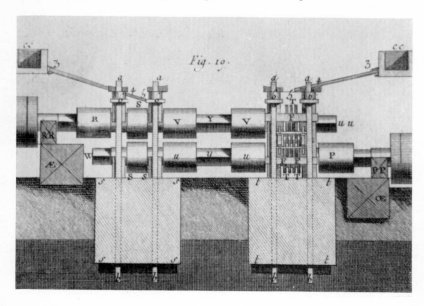

vided ore, timber for charcoal making, and reasonable streams for power purposes. This Weald area also had the advantage of being fairly near to a good market for iron—London. The supremacy of the Weald in ironmaking was to last a long time, until, in fact the latter part of the eighteenth century, when new techniques again caused fundamental changes.

It should be stressed here that the introduction of the blast furnace and its concomitant finery though having a far-reaching effect on ironmaking, took time to do so. There was no sudden or dramatic change. A few blast furnaces appeared, and it is unlikely that a contemporary observer would have any idea of their profound significance. It is only now, looking back, that we can assess the importance of this development and see it as the start of a new era. Industrialization had begun, feebly enough at first, it is true, but the pace was to quicken as time went on.

The blast furnace was a substantial masonry structure lined with fire-resisting stone and having an internal shape square in plan and in section not unlike two truncated square pyramids placed base to base, the lower, much smaller one, which was upside down, being placed over and opening into a short roughly parallel portion. These distinct (though interconnected) parts of the blast furnace eventually acquired names, which they retain today. They will be referred to many times in what follows, and they are therefore introduced here.

Starting from the bottom, there was first of all the hearth, a large block of fire-resisting stone laid flat. On this was built the parallel part of the furnace; this is known as the well or crucible; sometimes the word 'hearth' is used for the whole of the 'hearth and crucible'. Above the crucible came the inverted truncated pyramid or boshes; this has been fairly described as 'Like the hopper of a [flour] mill'. Lastly there was the second truncated pyramid, the stack of the furnace. Its open top was called the throat. There are no detailed records of the dimensions of the early blast furnaces, but the well would be about 3 ft deep, the boshes a little less, and the stack 12 to 15 ft high. Thus the total height would be around 20 ft, and the structure would be 18 or 20 ft square at the ground level, tapering by a foot or so to the top. The actual hearth was very small in plan, certainly no more than 1 ft square and probably less.

Archways at the base of the furnace structure gave access to the tuyere, through which the air blast was blown, and to the forepart, where the molten iron was drawn off or tapped at regular intervals. In time a second tuyere was added, as the furnaces grew bigger, then a third, and so on; today there may be twenty-six or more tuyeres in a single blast furnace.

Water-driven bellows kept up a more or less continuous blast at the tuyere, and once the furnace was lighted or blown in, it ran continuously. In theory it can go on running indefinitely, but in practice it has to be stopped from time to time for repairs to be made to the interior (or as it is now known, the lining). From the early days until charcoal was superseded by mineral fuel, the furnace actually stopped (or was blown out) much more frequently than repairs would have made necessary. If repairs alone were the criterion, the period in blast could extend to years. At a much later date a twenty-year period in blast was not uncommon and more than thirty years not unknown. In passing, it might be noted that present day practice, which is vastly more efficient but demands much more of the furnace lining, has reduced the period in blast once more, and a run of four or five years is customary.

In the days of charcoal firing it was unlikely that a furnace would remain in blast for a whole year. This was because a blast furnace, even a small one, used quite a lot of fuel, and the stocks of charcoal were soon exhausted, so the furnace had to be blown out. Sometimes, in the summer, the water supply would fail, and the furnace would have to stop. There followed a period of idleness for the furnace and fresh stocks of charcoal were accumulated. Then the furnace would be blown in again for another run; this period in blast was called a campaign, and it still is.

To blow in a blast furnace a charcoal fire was lighted on the hearth stone, a gentle blast was put on, and the fire was built up gradually. To get it going properly took at least some days, for the blowing in was always done very gently, to avoid damage to the furnace by heating it up too rapidly. Charcoal was added by tipping it through the throat of the furnace, and then a little iron ore was charged in the same way. The proportion of ore was increased gradually to the maximum which experience showed to be correct, and the furnace was then said to be producing normally.

Inside the furnace the carbon of the charcoal burnt in two stages, the reactions being shown chemically by the equations

$$2C + O_2 = 2CO$$
$$\text{(carbon)} + \text{(oxygen)} = \text{(carbon monoxide)}$$

$$2CO + O_2 = 2CO_2$$
$$\text{(carbon monoxide)} + \text{(oxygen)} = \text{(carbon dioxide)}$$

These reactions generate considerable heat, and the iron ore in the charge is raised to a point at which it is reduced by the reaction

$$FeO + C = Fe + CO$$
$$\text{(iron oxide)} + \text{(carbon)} = \text{(iron)} + \text{(carbon monoxide)}.$$

In actual fact the reactions taking place in a blast furnace are more complicated but the foregoing states in the simplest possible terms what happens. The metallic iron melts as a result of the high temperature in the furnace, and trickles through the burning charcoal to collect in a molten pool in the crucible.

Unfortunately for the blast furnace operator, all iron ores contain varying degrees of impurities such as sand, clay and other earthy matter, and some means of disposing of these, too, has to be employed. They will, in fact, combine with lime to form a waste material known as slag which, in the furnace, is also molten. Sometimes the iron ore has enough lime mixed with it naturally for the slagmaking reaction to take place; more often it has not, so a relatively small amount of limestone is also charged into the furnace to unite with the impurities, or act as a flux.

As long as the furnace is kept in blast, and is supplied with the necessary raw materials, these reactions continue. Iron and slag are produced, and waste gas (principally carbon monoxide) is given off. In the days of charcoal firing neither the gas nor the slag was used. The gas was allowed to burn at the furnace throat, and the slag was cooled and dumped. Today both are used, but the industry had to wait a long time before this was so.

Iron, being the heavier of the two molten materials, collected at the bottom of the furnace, and the slag floated on top of the iron. From time to time (originally every twelve hours or so, and now about every five or six hours) sufficient iron had accumulated for it to be taken off. First the slag was drawn off by opening a small hole (the slag notch) in the front (or forepart) of the furnace at the appropriate height. Then a second hole (the taphole) right at the bottom of the furnace well or crucible was opened, by knocking out the clay plug which stopped it up, and the molten iron was allowed to run out.

Initially the molten iron was run into a depression in a sand bed at the front of the furnace and there allowed to cool before it was broken up and subjected to the next process, for converting it to wrought iron.

This was an acceptable arrangement as long as the amount of iron tapped from the furnace was small; it must be remembered that it had to be handled manually. As the size of the furnace, and therefore its output, increased, even moderately, the stage was soon reached when a single piece of iron became unwieldy. So the practice developed of making a channel in the sand bed straight out from the furnace, putting side channels at right angles to it, and then making a number of further, dead-ended channels from the side ones. This resulted in a series of comblike moulds for the molten iron to run into, and when the iron cooled it could be broken easily into pieces of handy size.

The first channel, leading from the furnace, was called the runner, the side channels the sows, and the numerous small dead-ended channels, which were fed by the sows, the pigs. The allusion is an obvious one. The sand bed in which the sow and pig shapes were moulded, by pressing in wooden patterns by hand, and by cutting as required with a spade, was called the pig bed. All these names; runner, sow, pig and pig bed, persisted until recent years, when new methods of disposing of the molten iron rendered the big bed obsolete. But the term runner survives, and the pieces of iron, though no longer made in a pig bed, are still called pigs, or pig iron.

So far, our account of the indirect process has only dealt with the first part, the production of cast iron. It was not long before restricted uses were found for it in this form, but the main requirement was still (and remained for many years) for wrought iron. Conversion of the cast iron to wrought iron involved decarburizing it. This was done in a separate, charcoal-fired hearth, resembling a blacksmith's hearth and called a finery (not to be confused with the refinery, which is chronologically later, and will be dealt with in its proper place). The finery was different in one respect from the blacksmith's hearth, however. It had an air blast for the fire, and a secondary air blast which was played on the iron while it was being heated. Pieces of cast iron placed on the hearth soon became hot,

IRONMAKING UP TO 1700

and they were then stirred about with an iron bar and the secondary blast was turned on. The oxygen in the air blast combined with the carbon in the iron, which was gradually decarburized and so converted to wrought iron. Water-driven bellows provided the blast for the finery.

Now that the scale of production was bigger, machinery began to be used for working the iron as well as for providing the air blast for the furnaces. It was not long before the water-powered hammer began to be used in the forges, superseding the hand hammers of earlier times. There are records, in fact, that the water power hammer was used even before 1500, but it was a rarity then. After 1500, with the spread of the blast furnace and finery, use of the water-powered hammer increased accordingly. Quantities of iron produced were still very small, but they were getting beyond the unaided muscular power of man to handle. Thus, by about 1540, blooms of iron were made weighing a little over 2 cwt each.

The earliest form of power hammer was the tilt hammer and it is important to distinguish it from the helve hammer, which was also power driven, but came on the scene much later; the two are often confused. The tilt hammer was simple but effective, and it remained in use, in a few places, until recent years, though latterly for forging iron, not for working it in the sense meant here.

It consisted of an iron hammer head fixed on a wooden shaft (sometimes, confusingly, called the helve). This shaft was pivoted at or near the centre, and so fixed in its pivots that the hammer head worked over an iron anvil. The tail of the shaft was bound with iron bands, and a drum, with a number of projections or cams on its circumference, was rotated by a water wheel so that its cams struck the hammer tail in succession and released it. Each time a cam came in contact with the tail of the shaft, it depressed it, so lifting the hammer at the opposite end. A power hammer of this type, if operated slowly, would fall by gravity when the tail was released by the cam. But it would need to be quite heavy to deliver an effective blow, and this would naturally increase the overall size of the machine and the power required to drive it.

In practice, the tilt hammer was provided with a springy shaft and above and parallel with this shaft was a wooden spring beam. As the hammer was lifted by the cam it struck the spring beam and bent it momentarily upwards; the spring beam then drove it downwards faster than gravity alone would have done and as a consequence the hammer delivered a fairly rapid series of moderately heavy blows. The hammer head did not then need to be very large; a hundredweight or so would suffice. The spring beam was the common device for ensuring rapid blows, but some hammers dispensed with it and used a recoil block instead. This was an iron block fixed in the floor in a position in which it was struck by the tail of the hammer shaft as it descended; the result was the same as with a spring beam.

Because the available streams were often small, it was not uncommon for the different parts of a complete ironworks to be located at some distance apart, so that the same water could be used more than once. This was inconvenient in some ways, as it meant that pieces of iron had to be carried over a distance of perhaps a mile or more between processes, but it was sometimes unavoidable. This fact explains why old records often show that the blast furnace and the wrought iron works, or forge, as it was called, were separate, though they may have been in the same ownership.

Because the bloom of iron now being made was much larger than those produced in the old bloomeries, a new reheating device was needed for the forging process. This was the chafery, a charcoal-fired hearth very much like the finery, but lacking the secondary air blast. No change in the iron was made or intended in the chafery; it was solely for reheating the metal for forging to the shape required by the market.

Before the sixteenth century was out quite a number of ironworks in Britain were equipped with the three main components of the indirect ironmaking process—the blast furnace, the finery and the chafery and waterpower hammer, and the operating practice was pretty well standard. Pig iron from the furnace was decarburized in the finery to form a piece then known as a loop, and beaten out under the power hammer to make what was called a half-bloom, which was rectangular in shape. In this form it was of no commercial use, and it was next reheated in the chafery and again hammered to form a bloom.

The bloom at that time (the name has persisted to the present day and now has quite a different meaning), was made in what might seem, at first sight, to be a peculiar way. It was first heated in the middle and hammered down to a square bar, leaving a square knob at each end. These were called the mocket head (this was the larger of the two) and the ancony (the smaller). Further reheatings enabled the ancony and mocket head to be hammered down to the shape of the middle. This arrangement had two advantages. Firstly, slag was driven away from the centre of the bar towards the ends, where ultimately the surplus would be driven out altogether; secondly, as the bar gets longer it is much easier to thrust the ends into the fire for reheating than to try to reheat the middle. Finished sizes varied, but a fair average would be a bar an inch or two square and up to 40 lb or so in weight.

The power hammer had its limitations. It could not forge bars smaller than about $\frac{3}{4}$ in square, because below this size they became too long, and too flexible when hot, and they cooled down too quickly. Yet there was a demand for much smaller bars, or, as they are correctly called, when they are less than about $\frac{1}{2}$ in square, rods. Nailmakers, for example, needed small rods; so did blacksmiths and many other craftsmen.

The answer to this problem of producing small rods lay in the slitting mill, which was introduced to Britain in about 1588, but originated in Flanders much earlier.

Its date of origin is not known, but there were certainly some slitting mills in the Liège district soon after 1500. In Britain the date can be established more accurately, for a British patent was granted to Bevis Bulmer in 1588, and two years later Godfrey Box had a slitting mill in use at Dartford, Kent. It can be noted in passing that the dramatized story which for long credited a Midlands ironmaster, Richard Foley, with the introduction of the slitting mill to Britain is quite unfounded. Foley was the first to use the slitting mill in his own Midlands district, but this was not until 1628, or thirty-eight years after Box's mill was put to work. So the story (or stories, for there are several versions) about Foley disguising himself as an itinerant musician and wandering on the Continent to discover the closely guarded secret of the mill must be dismissed as pure imagination.

The slitting mill was a most important invention, and its significance goes far beyond the fact that it solved the immediate problem of making small iron rods. It was the first piece of true machinery after the power hammer to be introduced to the iron trade and, of even greater importance, it contained the elements of the rolling mill, which was to develop into a vital tool of the ironmaster.

Like many other good mechanisms the slitting mill was essentially simple. Two strong iron shafts were mounted one vertically above the other in an iron framework or housings and so arranged with rudimentary gearing that they could be turned simultaneously in contrary directions by a waterwheel. On these shafts were fixed discs or rotary cutters the edges of which intersected with each other in such a way that if a strip of iron were pushed against them it was drawn in and slit into pieces. Thus, if the rotary cutters were $\frac{1}{8}$ in wide, a strip of iron passed through the slitting mill would be slit into a number of narrow strips each $\frac{1}{8}$ in wide. If the original strip were $\frac{1}{8}$ in. thick, then the slit pieces would be $\frac{1}{8}$ in by $\frac{1}{8}$ in, a suitable size for nailmakers, who would round up the material for the nail shanks and forge heads as required, cutting each nail off the long strip as it was formed.

It will be noted that the raw material for the slitting mill was a strip of iron. This itself was not easy to make under the hammer, especially if it was thin. So the slitting mill had an extra feature, and a most important one it was. In a separate housing, in line with the slitting discs, and driven by the same waterwheel, were two plain cylindrical iron rolls. A piece of iron hammered out into a rough bar and then passed between these rolls was flattened and so prepared for the subsequent slitting operation quicker and far more effectively than could be done under the hammer. Knowledge of the fact that passing a piece of metal between rolls would flatten and elongate it was not new, but this was the first time it had been applied to the iron trade, and it marked a very significant step forward. Today the rolling mill is a most important piece of steelmaking plant, and it is worth noting that the slitting mill itself is still used, in a modern form, for trimming the edges of rolled steel strip and slitting it into narrower widths as required by the market.

With new methods and machinery, all of which worked on a larger scale than the old bloomeries, the industry naturally began to expand. Its expansion was slow, of course, but it was steady. By the year 1600 there were about eighty-five charcoal blast furnaces spread round the country and up to four or five forges for every furnace. This increase in numbers of works and in iron production meant, naturally, that much more charcoal was needed, and in this fact lay a danger which, by the early 1600s was serious, though it had been recognized earlier. The blast furnace, with its increased need for fuel, merely aggravated the problem.

Although there were extensive woodlands suitable for charcoal burning in the neighbourhood of the ironmaking centres, the supplies of timber were far from inexhaustible, and the ironmakers were not the only ones who wanted timber. In particular the builders of ships, especially naval ships, were recognized as having a legitimate claim on the timber reserves. There were several attempts to protect the forests by legislation (most of them being wholly ineffective) and some success was achieved in timber conservation by replanting and in the growing of coppices specifically for charcoal burning.

All these measures, however, even when they were successful, which they were not invariably, could only provide a part answer to the problem. At best they might have succeeded in holding the iron trade at a steady level. They could not, and in fact they did not, allow it to expand to any really important degree. The answer lay in the substitution of some other fuel for charcoal. This was seen to be so by a number of people, and the alternative fuel, coal, was named in a number of patents taken out in the seventeenth century. None of these patents was successfully applied, and the names of most of the patentees have passed into history without comment. It was one thing to propose the use of coal as a fuel and another matter altogether to find a way to do it, for although it seems, superficially, quite easy, it proved to be very difficult.

Coal burns, like charcoal, and a major constituent of it, as of charcoal, is carbon. But coal contains numerous impurities, notably sulphur which, as has been pointed out, has a disastrous effect on the quality of wrought iron. Moreover, though this was not so important at first, coal does not burn so freely in masses as does charcoal, and a furnace fired on coal alone would need modifying as regards internal shape or lines, and would need different blast arrangements. So it is not surprising that the early projectors found, when they tried using coal (and there is no doubt that some of them did try) that it did not give the results they had hoped for. It was not until early in the eighteenth century that a means of using coal successfully in the blast furnace was found. That discovery will be discussed in the next chapter.

However, before leaving the seventeenth century coal fuel proponents, the name of one must be mentioned. He is Dud Dudley, and so much has been written about him in the past that something must be said of his work here, if only to put the matter in proper perspective.

Dud Dudley (1599–1684) was the natural son of Lord Dudley, on whose large estates in the Midlands there were several ironworks. Dud Dudley managed these works for his father and was unquestionably a practical ironmaker. His principal claim to fame lies in the fact that he wrote a book *Metallum Martis* in 1665 and in it he claimed that he had succeeded many years earlier, in making good marketable iron with coal fuel in one of his father's furnaces. This claim has been accepted uncritically and repeated by generations of writers.

Certainly two patents were granted for Dudley's process. One, dated 1621, was in the name of Lord Dudley, the other granted in 1638, was a joint patent of Dud Dudley and three partners. Unfortunately neither the patents (as was customary at the time) nor *Metallum Martis*, give any technical information at all. The only evidence that can be gleaned is that Dudley implies that he used the slack or small coal from the Staffordshire Thick, or Ten Yard coal as his fuel, pointing out that this was a low-price commodity, as it normally went to waste (this was true at the time).

But the Ten Yard coal contains an appreciable amount of sulphur, which would not have been removed in the subsequent working into wrought iron, and it would therefore be impossible to make a consistently good wrought product.

Eventually, the problem of using coal as a fuel for the blast furnace was solved by coking the coal, but Dud Dudley makes no claim to having coked his fuel, nor could he have done so, for the Ten Yard coal is non-caking, that is, it is not of a type suitable for coke-making. Eventually, too, it became possible to use raw coal in the blast furnace, and this was done in a few places, but only after developments which took place subsequent to the 1830s. So Dudley could not have used the raw coal he mentioned, and he could not have coked the coal first.

Dudley may have made a bloom or two of satisfactory wrought iron from coal-smelted iron, by good fortune, but he could not have done so consistently. In fact he admits as much, though he blames his failure on the machinations of his fellow ironmasters and on to the Civil War, in which he suffered greatly on account of his Royalist sympathies. Nevertheless, though Dud Dudley's claims do not stand in the light of technical investigation, he deserves credit for having seen the problem so clearly and for having tried so hard to solve it.

CHAPTER THREE

New processes and the Spread of Mechanization 1700 - 1800

The man who found the answer to the charcoal problem was Abraham Darby (1676–1717) a Quaker ironmaster who was born near Dudley, learnt the trade of ironfounder in Bristol and in 1708 took on lease a little charcoal blast furnace at Coalbrookdale, Shropshire. His object in taking the furnace was to have control over his own supplies of iron for his foundry, for the making of cast iron pots was his main business. The foundry at Coalbrookdale was to extend its activities into many branches of ironmaking and to achieve world fame under Darby and his successors, several of whom had the same Christian name; for this reason the successful mineral fuel pioneer is usually known as Abraham Darby I. The foundry is still carried on at Coalbrookdale, on almost the same site, and an excellent museum, showing the development of the business and its products over the years is maintained there by the present owners, Allied Ironfounders Ltd.

But when Abraham Darby I took over the furnace, it was very small and of no particular importance. What the site did have, however, was all the necessary raw materials for expansion and development. Timber was available, together with ironstone, clay, sand, limestone and, what is more important, coal, all at a short distance from the furnace. A little stream, suitable for powering the furnace bellows, ran down the valley through the works to join the river Severn, and the latter river provided a useful highway for the despatch of finished goods.

Darby evidently lost no time in getting to work on the business of using mineral fuel, for records still in existence show that in 1709, the year following his arrival at Coalbrookdale, he was smelting iron with coal. Incidentally, though it has long been recognized that

Abraham Darby I was the first person to succeed with mineral fuel, various dates for his historic success have been given in the past. There is no doubt, however, that 1709 is correct; detailed researches in the last twenty years or so have established this fact beyond question.

Darby succeeded where others had failed because he coked the coal he used as fuel, instead of attempting to use it raw. He did not invent the idea of coking coal; coke was already in limited use in his time for such purposes as malting, where the smoke and sulphur fumes from raw coal would be unacceptable. It is significant that Darby was apprenticed to a malt-mill maker, and so had every reason to know of the use of coke in malting. Coke was made in the same way as charcoal, by a process of controlled burning in heaps under a layer of wet ashes until the smoke and fumes had ceased, indicating that only carbon remained. The heap was then broken up and the glowing mass was quenched with water. What remained was almost pure carbon, very similar to charcoal. The sulphur and some of the other impurities in the raw coal had been driven off. This is still, fundamentally, the method of coking today, though fixed structures are used instead of ash-covered piles, the work is mechanized, and full use is made of the gases which are given off during the process.

In Darby's time and for long after no attempt was made to use the gases. What Darby wanted was coke. By good fortune the local Shropshire coal had a low sulphur content and by coking it Darby got rid of most of what it did contain. This coke proved to be a very satisfactory fuel indeed, though the solution to the problem of using it in place of charcoal was not quite as simple as it might seem. Darby left no records showing what technical difficulties he had to overcome, but it is not difficult to suggest what some of them might have been.

Coke does not burn quite so readily as charcoal, and Darby would find that he needed to provide a stronger blast and a greater volume of air. Hence, he would have to modify his furnace bellows. Again, he would probably have to make some alteration to his furnace, for the internal shape or lines of a blast furnace is quite critical in detail, and different fuels call for small but significant changes. Darby would also have to work out, by experiment, the correct proportioning of ore and coke, or, as it is called, the burdening of the furnace (the burden is the ratio of coke to ore) and he would find that coke is capable of carrying a heavier burden than charcoal, since it does not crush so easily. This latter consideration is really of a long-term nature; furnaces did, in fact, eventually begin to increase in size as a direct result of the use of coke fuel. The other considerations were of more immediate interest.

Though he left no notes of his technical problems, it is known that Darby did have to experiment before he was successful. It is recorded that he spent a lot of time at his furnace, even, on occasion, whole days at a time, snatching what sleep he could on the spot. It has always been, and it remains, a characteristic of a blast furnace, that it takes time to see what effect an alteration in the method of working has. The test is, does it make good iron consistently? To find out, the experimenter must wait until the furnace has been tapped several times to see what effect his ideas have had. If the desired result has not been achieved, he must try again—and wait again. There was no way of hurrying things up. This is true to some extent even today, though there are ways now of calculating many factors in blast furnace operation with considerable accuracy. In Darby's time the only approach was that of trial and error.

That Darby succeeded in using coke made from the local coal as a substitute for charcoal in his blast furnace is beyond question. His success is one of the great landmarks in the history of the iron trade, for it freed the ironmakers from the restrictions of the charcoal famine, and put them a step forward in the great expansion of the trade which was to follow. Again, though, as was the case with other fundamentally important discoveries, there was no sudden or dramatic change. Coke smelting was slow to spread, and even more than a century later there were still some charcoal furnaces in use.

There were several reasons for the slow expansion of coke smelting. Darby was an ironfounder; he wanted iron for his casting business and had no interest in wrought iron, which was still the main product of the trade. His successors at Coalbrookdale did, in fact, take up wrought iron manufacture, but Abraham Darby I confined himself to the foundry trade. He did not patent his process, but he made no attempts to publicize it either. Coalbrookdale at that time was rather remote, and the news took a long time to travel. Moreover, since Darby was not making wrought iron it is unlikely that other ironmasters, even if they heard of his method of smelting, paid much attention to it at first. Again, some areas where coke smelting would have been of interest were, at that time, prevented from using it by lack of water power. Such an area was the Black Country district of South Staffordshire, not far from Coalbrookdale. This area had everything it needed for ironmaking except water power, which was negligible. So it had to wait until another form of power was developed.

Nevertheless, coke smelting did spread to other parts of the country. By 1788 the number of charcoal furnaces in England and Wales had fallen to twenty-four, while the number of coke furnaces had risen to fifty-three, of which twenty-one, not surprisingly, were in Shropshire. There were a few coke furnaces in Scotland as well. Production had risen too; the average annual output from the charcoal furnaces in 1788 was 545 tons, while that of the coke furnaces was 909 tons.

Superficially a coke-fired blast furnace was very much like the older charcoal furnace. In fact, as Darby had shown, coke could be used in a charcoal furnace, and several besides Darby's were simply converted. Coke, however, lends itself to the construction of larger furnaces, and those which were built specifically for coke fuel were larger. In addition the coke furnace was able to run continuously, without having to blow out from time to time while supplies of fuel accumulated. One long-standing problem of the iron trade, that of fuel shortage, had been solved in part. It was only in part, for charcoal was still needed for the forges. This, too, was overcome in due course, before the eighteenth century was out, as will be seen.

A more immediate problem, following Darby's success with coke, was that of power. Water power is limited in this country, and suitable streams were not always where they were needed. Furthermore, they are liable to freeze in winter and dry up in summer. This need not be such a serious matter with charcoal furnaces. A period of water shortage could be used to accumulate fuel. A coke furnace, on the other hand, needs plenty of power all the year round. The answer lay, of course, in the use of steam power, and it was fortunate that while Abraham Darby I was at work on his epoch-making discovery, the first steps were being taken to develop a reliable source of power which would be independent of the uncertainties of streams and rivers.

The story of the development of steam power is dealt with in another book in the present series, and it is only necessary to give here a bare summary of the events, which led up to a vast expansion of the iron trade. Savery's engine of 1698 was of no use to the iron trade, and Thomas Newcomen's, which followed in 1712, was really only of direct use for mining. Newcomen's engine, however, did have a connection with the iron trade, for the Coalbrookdale works supplied a number of cast iron cylinders for Newcomen engines, thus marking the beginnings of a connection between the iron and engineering trades which grew in importance as time went on, and remains vital today. And Newcomen's engine, which was the first to have a cylinder, piston and rocking beam, paved the way to the inventions of James Watt, which were of real importance to the ironmakers.

Watt's engine, at first capable of reciprocating motion only, was, like Newcomen's, of major interest to the mining industry. Nevertheless, while the first engine made by the Boulton and Watt Soho partnership of 1775 went to a colliery for pumping, the second was supplied to an ironworks. This second engine was designed to the order of John Wilkinson (1728–1808), who was probably the most famous of all the eighteenth-century ironmasters. It was used to operate a blowing cylinder for Wilkinson's blast furnace at Broseley, Shropshire, not far from the historic Coalbrookdale works. So far the Watt engine could not power rotating machinery—though it was later adapted to do so—but if it could work a reciprocating pump for water, it could equally well work a reciprocating pump for air for blowing a furnace, and that was the job Wilkinson gave it to do.

In this particular instance the engine was designed by Watt, incorporating his patented separate condenser, and built by Wilkinson. This was a special arrangement between Watt and the ironmaster, 'Iron-mad Wilkinson' as he was often called because of his outstanding enthusiasm for iron. Wilkinson was for many years the supplier of the cast iron cylinders and pistons for Watt engines, casting and machining them to the orders of Boulton and Watt, and despatching them direct to the place of installation, where they were erected, with the other components, under the supervision of Watt or a Soho engineer.

It so happened that Wilkinson was not only a large-scale ironmaster, but, at the time when Boulton and Watt entered into partnership, the only person who could bore cylinders to the degree

5 John Wilkinson's Bradley works, Bilston, Staffs

6 Brymbo No. 1 furnace. An 18th century masonry blast furnace as it was at the end of the 19th century

7 Three 19th century blast furnaces charged from bank at rear, at Blist's Hill, Shropshire

8 Three 19th century blast furnaces at Priorfields in Staffordshire, charged by hoist in tower on the left

of accuracy needed by the Watt design. Watt's engine called for much more precise machining than had been necessary with the Newcomen engine, and it was achieved with a machine invented by Wilkinson in 1774 for boring cannon, and easily adapted to deal with engine cylinders. Wilkinson's boring machine was essentially simple, but it incorporated for the first time a rigid boring bar which ensured that the bore of a cannon, or a steam engine cylinder, was not only truly round, but truly cylindrical as well.

The business association of Wilkinson and the Boulton and Watt partnership was a fruitful one, and emphasizes the growing interdependence of the ironmasters and the engineers. In his own business of ironmaking Wilkinson's enthusiasm for innovation knew no bounds. He operated blast furnaces and ironworks in Shropshire, Staffordshire, North Wales and elsewhere, and was a pioneer in the use of iron for many purposes which were, at the time, unusual. He built the world's first iron boat in 1787 and launched it on the river Severn, where it promptly confounded his many critics by floating successfully; he built a chapel for his workpeople at Bilston in Staffordshire and used iron for the door and window frames and for the pulpit (which still survives); and he even had a cast-iron coffin made for himself, though he was not buried in it, having made no allowance for his increase in girth over the years.

Wilkinson was also one of the promoters of the famous iron bridge which gave its name to the Severnside township of Ironbridge near Coalbrookdale, though he was not the builder of it. The bridge was primarily the work of Abraham Darby III (1750–89), the grandson of Abraham Darby I, who was then in control of the Coalbrookdale works. The bridge stands today, scheduled as an ancient monument, and is truly a remarkable tribute to the foresight and skill of the men who planned, designed, made and erected it, for it was the first of its kind in the world. There was no precedent for an iron bridge, and only the enthusiasm of a a few who had faith in iron, and in their own ability to make and fashion it, made the bridge possible. It has a span of 100 ft, a width of 24 ft, and a height of 45 ft above the river. Not only is it all of iron, but considerable ingenuity is shown in its design, for the main members were assembled by threading one through another and tightened in place by wedges. No nuts, bolts or rivets were used. A few nuts and bolts are to be found there now, it is true, but these were used in subsequent repair work to damage caused principally by movement of the ground on which the bridge abutments stand.

By the second half of the eighteenth century ironmaking had undergone great changes and was all ready, except in one respect, to start the great expansion which was to follow after about 1800. Darby's coke smelting process had freed it for good from the dangerous position into which the charcoal shortage had led it. James Watt's engine had given it a new and adaptable source of power, which enabled the industry to expand and to move into districts such as the Black Country, where there were abundant supplies of raw materials but insignificant water resources. And by 1781 Watt had adapted his engine to produce rotative motion, that is to turn machines of all kinds, besides the machines used in ironworks. Further improvements to Watt's design in 1784 made a better engineering job of the rotative engine, and improved its efficiency generally. These developments were of benefit to the ironworks in two ways. They provided a new source of power and, since the new engines were in great demand in other industries everywhere, they opened up new markets for the various iron castings and forgings which were needed in steam engines.

The manufacture of cast iron, and the working of wrought iron by power-driven machinery, however, had got out of step with the actual making of wrought iron itself. In this department the old finery was still in use. It was slow and inefficient, its output rate per unit of production was small, and it still needed charcoal as a fuel. Clearly there was a need for an improved method of making wrought iron from blast furnace pig iron.

As has happened so often before and since, the need was recognized by several people, and efforts were made to develop a process, more than one patent being taken out in this field.

It was Henry Cort (1740–1800) who succeeded in making wrought iron by a new method, but the brothers Thomas and George Cranage and Peter Onions deserve mention because they came so near to success before Cort actually made wrought iron using raw coal as his fuel. The Cranages patented their process in 1766 and Peter Onions was granted a patent in 1783. All three men came from Coalbrookdale, and both patents specified the use of a type of furnace which Cort used later. An examination of the patents shows that either of them could have worked, and it may be that one or both was actually put to work. But for reasons which are not now known, neither was ever brought into commercial use.

Henry Cort, working at a little forge at Fontley, in Hampshire, an unimportant ironmaking area, used a reverberatory furnace to melt down pig iron, and decarburized it to make wrought iron. The reverberatory furnace was not new, having been used for certain metallurgical puposes before Cort's time, but Cort was the first to employ it for making wrought iron. His experiments succeeded very well, and he patented his process in 1783 and 1784.

The reverberatory furnace was basically two large iron boxes, lined with firebrick, one of which (the smaller) was equipped with firebars on which there was a coal fire. The second, larger, box contained the iron to be decarburized, and the iron and fuel were separated by a wall or firebridge of firebrick, over the top of which the flames from the fuel passed. A flue at the opposite end of the larger part of the furnace was connected to a chimney stack, and the draught drew the flames across the iron in the furnace. As the top or roof of this part of the furnace sloped downwards towards the flue, the flames were deflected or reverberated down on to the iron; hence the name reverberatory furnace. Both parts of the furnace were fully enclosed except for an aperture at the front of the metal portion to give working access, and a similar but smaller aperture in the firebox end to enable it to be fuelled as required. A counterbalanced door kept the working aperture closed except when the furnace was being charged or emptied. For working the iron during the process a small subsidiary aperture in the working door sufficed. No door was provided on the firing hole; this was kept closed as far as was necessary by a lump of coal.

Cort's process enabled him to use raw coal as a fuel because, in the reverberatory furnace, the fuel and the iron being decarburized were physically separated by the furnace firebridge. If coal had been used in the finery, sulphur from the coal, which was in contact with the iron being decarburized, would have passed into the iron. Cort charged cold, solid pig iron into his reverberatory furnace and melted it down. As the iron melted, it was stirred about with an iron bar, to expose it as fully as possible to the heat, and to the current of atmospheric air which was drawn through the furnace by the chimney draught. This brought about the required reaction. Oxygen from the current of air united with the carbon in the iron, the product, carbon monoxide gas (CO) passing away through the flue and up the stack. In due course the carbon had all been 'burnt' out, and the iron which remained was fully decarburized, or wrought iron. Cort defined this period in the process by saying that the iron had 'Come to nature', and this term persisted right through the wrought iron period.

Because the molten iron was stirred by means of an iron bar to promote first melting and then decarburization, the process got the name of puddling, and this term, too, lasted as long as wrought iron itself, although it was eventually modified by the addition of the word 'dry', to distinguish it from the later wet puddling process; of this more will be said in the proper place. Puddling, incidentally, was quite hard physical work, especially as the iron was coming to nature, for it will be remembered that the higher the carbon content the lower the melting point. Conversely, as the carbon was removed, while the furnace temperature remained the same, the iron changed from a molten to a pasty condition, and got progressively stiffer to work.

Cort's puddling process was a success and it spread fairly rapidly, in spite of the fact that it had certain limitations. Not the least of these was the fact that it was really only capable of dealing effectively with a pig iron which was low in carbon and silicon. Using only atmospheric air as its decarburizing agent, Cort's process worked slowly on both the elements named, and to render it economic, it was necessary to start with what was called white iron, or low-carbon iron. This could be made in the blast furnace, though to do so was difficult, and it became the custom to treat the blast furnace iron in a refinery before charging it to the puddling furnace. The refinery, which was similar to the earlier finery, but fired with coke, produced the necessary white iron by partly decarburizing and desiliconizing the blast furnace metal. Cort's process was really, therefore, a double one, as it was practised.

Puddling was the one thing which was needed to fill the technological gap in the iron trade in the second half of the eighteenth century, and it became the standard method of wrought iron production, a position it retained for about half a century. It enabled the production of wrought iron to keep pace with the other, larger-scale processes already in use, and it made further expansion of the iron trade possible.

One other development by Cort must be mentioned, for it was of great importance. This was the use of grooved rolls for producing rolled shapes other than flat products. Cort covered this process in his patent of 1783. It has already been mentioned that the slitting mill had an auxiliary pair of rolls for producing flat strip iron for slitting. Another rolling process had also been in use since about 1720 when John Hanbury used plain rolls for making sheet iron for tinplate. This, it should be noted, is misnamed, for it was formerly thin sheet iron, and is now thin sheet steel, coated with tin; it is neither tin nor plate, but the name has now been used for so long that it is established beyond anyone's power to change it. Both the slitting mill rolls and those used for sheet rolling were plain cylinders.

The rolls used by Cort were different. By cutting suitably shaped grooves in the circumferences of his rolls he was able to roll bars. Thus, for example, if a groove of, say, $\frac{1}{2}$ in radius were cut in each roll, the pair of rolls would be capable of rolling an iron bar 1 in diameter. It is obvious that other shapes, such as squares, are equally feasible. Cort was not the originator of grooved rolls. They had, in fact, been both patented and used before his patent was granted, but the significance of his work is that he made a success of them.

So much had happened since Abraham Darby I succeeded in using coke as a blast furnace fuel that the whole picture of the iron trade had changed. When the eighteenth century began, ironmaking was certainly an industry, in that it used some heavy, fixed capital equipment and machinery, and some sections of it at least, operated continuously. But it was localized and not very impressive in the matter of output. No figures are available for 1700, but by 1788, when the effect of the coke furnace was making itself felt, the blast furnaces of Britain were producing something like 68,000 tons of iron a year. Just after the end of the century, in 1806, the figure was 258,000 tons.

It is therefore appropriate at this point to take a more detailed look at ironworks plant and machinery as it was towards the end of the eighteenth century. By that time the blast furnace had grown in size, though its general appearance and shape had changed very little since it was first introduced. Its growth had been due to two factors; coke fuel, and mechanization. Coke, as has been said, will support a greater weight than charcoal without crushing, and furnaces by say the 1780s were 1 or 2 ft square in the hearth, 5 to 6 ft in the boshes and 30 ft or more high. Dimensions varied quite a lot from place to place, but those given above are reasonably typical.

An important feature of the furnace was that not only had steam power been applied to blowing it (though not universally of course) but steam had also been used to ease and cheapen the work of charging. Early blast furnaces were charged with fuel, ore and limestone quite literally by hand, the materials being carried in baskets by workmen and tipped into the furnace throat. To get to the top of the furnace from ground level, where the materials were assembled, a ramp was built up in the form of a bridge, or earth bank, and up this the men walked. In some localities the furnace could be built against or into the side of, a hill, and then it was only necessary to build a short wooden bridge to give access to the furnace throat.

As furnaces got bigger this arrangement of carrying the materials became impracticable, and wheelbarrows were used instead. Then, when the rotative steam engine became available, from 1781 onwards, the practice grew of using an engine to haul the barrows up the incline. From this, in due course, developed the engine-hauled wheeled platform running on the incline, on which three or four barrows could be carried simultaneously. Sometimes two platforms were used, on parallel tracks on the incline, being interconnected by chain or rope so that they counterbalanced each other and the engine only had to overcome the resistance of the load. Sometimes there was only a single platform, with a counterbalance weight on wheels running beneath it, but the effect was the same. By the end of the century an engine-worked incline carrying barrows was the standard blast furnace charging equipment for most new furnaces; the exceptions were generally in places where the natural lie of the land made it easy to assemble the materials at furnace-top height and wheel them across a short, level bridge. Such an arrangement can easily be visualized from the remains of a furnace of 1777 preserved at Coalbrookdale.

For blowing the furnace the usual arrangement was either a double water-driven bellows (where conditions were favourable) or a double-acting steam-powered blowing cylinder. In either case the blast was practically continuous, for the twin bellows were set so that one emptied into the air blast pipe while the other filled, and the blowing cylinder was arranged to discharge air in both directions of travel.

At the lower part of the furnace, an archway in the side or back of the masonry gave access to the tuyere; wear and tear on tuyeres were heavy and they needed changing often. At the front of the furnace a similar arch gave access for tapping the iron and slag. This arch was known as the tymp arch, from the heavy cast iron beam or tymp which supported the masonry, and particularly the lining of the furnace over the inner part of the arch.

Right down at hearth level was a low wall, the dam, of heat-resisting stone, usually supported on the outside by an iron dam plate. Molten iron collected behind this dam as the ore was reduced in the furnace, and was drawn off or tapped every twelve hours or so, for which purpose there was a small hole, the tap hole, which was normally closed by a plug of clay. At the top of the dam, usually on one side of it, was the slag notch, a slot through which molten slag was drawn off as required.

Between the top of the dam and the underside of the tymp there was a space but this was normally kept closed by clay held in place by an iron plate propped in position from outside. This plate remained in position while the furnace was in blast, but at tapping times the blast was taken off for a time, the plate was removed, and the clay broken out, so that the furnace workers could gain access to the lower part of the furnace for examination and for the removal, if necessary, of any unburnt coke, unsmelted ore, and slag. For this purpose they used long iron hooks and a very long-handled iron shovel. When the iron and slag had been tapped and the hearth cleaned out as required, the hearth was sealed off with clay, the plate was put back, and the blast was turned on again. If all went well, there would be nothing further to do at this part (the forepart) of the furnace for another twelve hours, except to tap off some slag once or twice.

How often the slag was tapped depended, naturally, on how fast it was made, and this in turn was a matter of how pure or impure the iron ore was. The more impure the ore, the greater the quantity of waste there was to be taken off as slag. Thus, the number of slaggings varied from place to place; two in between tappings and a final one just before tapping, known as the casting flush, would not be unusual. Slagging was generally so called, but it was also known as flushing and the slagger or man who did the actual job of slagging was often called the teazer. To tap slag it was not necessary to remove the clay stopping and the plate under the tymp. All that had to be done was to open the slag notch, which was usually kept closed by an iron bar with a 'head' of slag on it.

This, then, was the typical 'new' or modern blast furnace plant at the end of the eighteenth century. Such plants were common, but a few of the older ones remained in use. In Sussex, for example, though the area had dwindled in importance from its former leading position to a producer of no real significance, one or two charcoal furnaces remained in blast. The last furnaces there, at Ashburnham, carried on in fact until 1809, or 1810, and the last forge, at the same place, closed down in 1820. Elsewhere, too, a few charcoal furnaces carried on, and one actually lasted until about 1917, but charcoal furnaces had had their day. Charcoal iron was no longer of importance.

At the forges the picture was much the same as at the blast furnaces. While the old fineries and chaferies had by no means disappeared by 1800, they were dying out, and new works used the new puddling process and, generally, steam power. So a typical ironworks of the period might have one or more blast furnaces, refineries, puddling furnaces, power hammers and rolling mills. No longer was it necessary for the works to be where the water power was, and to be spread out along a stream to make the best of it.

The day of the integrated works had begun, with all the producing units close together and conveniently situated for supplies of raw materials. Here again, though, it should be noted that not every works was fully integrated. Some ironmasters made only pig iron and sold it to ironfounders and to forgemasters who did not own blast furnaces. This pattern has continued to the present day; not every works carries out all the processes necessary to make its finished product. But by the end of the eighteenth century the facilities existed for those who wished to run integrated plants, and some quite large concerns were already in existence. The celebrated Coalbrookdale Company was just such a concern. It still had its blast furnaces and its foundries, and still made the cast iron hollow-ware which was the reason for its founding by Abraham Darby I. It had branched out considerably, however, and had several works in the Coalbrookdale area, including engineering works and wrought iron works. Besides this, the company mined its own raw materials and had an extensive system of narrow-gauge tramways connecting its various works, pits and river wharves. It was not alone; there were others like it in various parts of the country, though a cross section of the industry would have disclosed the small as well as the large and this pattern continued throughout the next century and right down to the present day.

CHAPTER FOUR

Wrought Iron Ascendant
1800 - 1860

In the first half of the nineteenth century there was a great expansion in all branches of ironmaking, though wrought iron continued to be of major importance. Against a background of ever-increasing industrialization, which provided an expanding market for it, wrought iron developed, with the aid of new manufacturing techniques, to reach its heyday in about the middle of the century. Up to about 1860 there was nothing which could offer a serious challenge to its position as the supreme ferrous metal. By the 1870s a serious competitor, mild steel, was making its presence felt, and although the older metal had a lot of life left in it, no further significant technical advances took place in wrought iron manufacture, and its fate was slow decline.

However, although wrought iron held pride of place up to the 1860s, the other main department of ironmaking, the blast furnace, was far from stagnant. It is convenient to consider developments in that field first, since the product of the blast furnace still went mainly to the wrought iron forges, and improved methods of pig iron production were the basis of expansion in wrought iron.

For some years into the nineteenth century, there was nothing of importance to report in blast furnace operation, though the subject was beginning, for the first time, to receive the attention of what we should now call scientific investigators. Prominent in this field was David Mushet (1772–1847), a Scot, who wrote a series of papers for the *Philosophical Magazine* on iron and steel. These papers were later collected into a substantial volume (*Papers on Iron and Steel*, 1840), and Mushet achieved wide acclaim as both a practical and a scientific ironmaker. He was also the father of the even more famous Robert Forester Mushet (1811–91) of whose achievements more will be said at the appropriate point in this chronicle.

WROUGHT IRON ASCENDANT, 1800-1860

David Mushet is probably best known for his discovery of the Blackband ironstone of Scotland. This ore, discovered in 1801, got its name from the fact that it contained a proportion of coal. It was used in great quantities in Scottish ironworks, and although it is now worked out, it was responsible for considerable expansion in Scottish ironmaking; many Scottish ironworks owed their prosperity to Blackband ore. Mushet moved from Scotland to Derbyshire and then, in 1810, to Coleford, in the Forest of Dean, where he remained for the rest of his life, continuing his experiments, and having a financial interest in a small ironworks.

While he was still in Scotland, however, Mushet came into contact with another Scot, James Beaumont Neilson (1792–1865), whose work on the blast furnace was to have far-reaching effects. Neilson was the son of Walter Neilson, who was then in partnership with Mushet, and although Mushet had actually left Scotland before Neilson's invention, of hot blast for the blast furnace, was announced, he knew and approved of it, and gave evidence in its favour later, at a patent infringement trial.

J. B. Neilson had not followed his father into the iron trade. He was manager of Glasgow gasworks when, in 1828, he patented hot blast. But he understood blast furnaces, and he was given facilities locally to try out his idea. The invention was simple and logical and though as is usually the case with novel ideas, there were many who decried it. These included some ironmasters who were ready enough, in the end, to try to use it without the patentee's permission.

What Neilson did, simply, was to heat the air blast before it was blown through the tuyeres into the blast furnace. It was an important invention which gave immediate economies in blast furnace operation. These economies were improved as the blast-heating apparatus itself was developed, and hot blast ultimately became universal. All modern blast furnaces are blown with hot blast. But when Neilson first put forward the idea of heating the blast, few ironmasters viewed it with favour, for at that time it was generally considered that the colder the blast the better. So any schemes for heating it were bound to be suspect.

Those who supported the idea of cold blast based their arguments on a faulty interpretation of the fact that a blast furnace often seemed to work better in cold weather than in hot. Because of this fact a few experiments were actually made to cool the blast by passing it over or through water. In fact the variation in air temperature had nothing to do with it. What really caused fluctuations in the production of the blast furnace was differences in natural humidity. In warm humid conditions the air contained an appreciable amount of moisture, while in cold dry conditions the water content was much lower. Water vapour, H_2O, is of no use to a blast furnace, which needs oxygen. So when the humidity was high the correct thing to do would have been to blow more air into the furnace. This, being unknown, was not done, and the production variations were attributed to the air temperature.

Naturally, when Neilson proposed to do the exact opposite of what some experienced ironmasters considered correct, there was scepticism. But experiments in Scotland showed that heating the blast certainly did have beneficial effects, and in due course the process began to be adopted. Neilson showed that where a total of 8 tons $1\frac{1}{4}$ cwt of coal had been used to make a ton of iron with cold blast, if the temperature was heated to about 150°C, the fuel needed was only 5 tons $3\frac{1}{4}$ cwt (in both cases the weight of fuel was that of the raw coal converted into coke—the furnace was, of course, at that time coke fired). When the temperature of the blast air was raised further, even greater economies resulted.

In Neilson's time the problems of heating the air were considerable. His first air-heating device, which produced an air temperature of about 90°C, was simply a chamber made of wrought iron plates, with a coal fire underneath it. A pipe led into one end from the blowing engine and a second pipe, at the opposite end, conducted the heated air to the furnace blast main and so to the tuyeres. This heater was not very successful, for the iron plates of which it was made were soon burnt through by the fire. So Neilson made his blast heaters of cast iron, which resists the action of fire much better than wrought iron plate, and using one cast iron vessel to each tuyere, raised the blast temperature to the region of 138°C to 150°C. In due course, with improved stoves the blast temperature was lifted to about 315°C, and there it remained until a completely new type of stove was introduced several years later. This was the Cowper stove, which will be dealt with in the appropriate place. It need only be said here that hot blast, once accepted, had come to stay. Today, incidentally, blast temperatures of 1000°C or more are common.

An important incidental effect of the hot blast was its development of the need for a new type of furnace tuyere, and a new method of connecting it to the pipe, or blast main, which conveyed the air from the blowing engine to the furnace. It had always been necessary to provide means for changing tuyeres while the furnace was at work, for tuyeres wear and are damaged easily. With cold blast the tuyere was connected to the blast main by a leather pipe, known as a bag. This could be disconnected quickly for tuyere changing. When the blast was hot, even at the low temperatures used at first, the leather bag was obviously unsuitable, and quick-release cast iron jointed fittings, known today as goosenecks, were devised instead.

It also became necessary to provide some means of cooling the tuyere itself, for not only was hot air passing through it, but the temperature in the furnace well was also greater than in the cold blast furnace. The answer to this problem was provided by water cooling the tuyeres. Water-cooled tuyeres were not unknown in Neilson's time, but they were not common; hot blast made them, in time, universal for blast furnaces. Water cooling of tuyeres was done in one of two ways. Either the water was circulated through a coil of wrought iron pipe embedded in a cast iron cone, or it passed through the annular space between two concentric, truncated cones of wrought iron plate. The first, Condie's tuyere, was also known as the Scotch tuyere, while the second, of wrought iron plate was called, from its place of origin, the Staffordshire tuyere. It is basically the Staffordshire tuyere which is used today, but instead of being made of wrought iron plate it is a copper casting.

A further effect of hot blast, though not of much importance at first and never of very great significance, was that it ultimately made it possible to use raw coal for smelting iron. This, as has been stressed, was impossible in the older furnaces as the sulphur in the coal made the final product, wrought iron, hot short. But when temperatures were raised sufficiently by means of the hot blast, it became possible to remove the sulphur, using a suitable flux, as calcium sulphate in the slag. Raw coal smelting never assumed great proportions, but it was certainly practised in Scotland and North Staffordshire and a few other localities where conditions were favourable for it.

Not long after Neilson patented hot blast another development took place in blast furnace operation, which was also to have far-reaching effects. This was the change in the internal form, or lines, of the blast furnace, introduced by John Gibbons, from the Black Country area of Staffordshire, in 1832. It had long been recognized that the internal shape of a blast furnace could have a marked effect on its operation, and many minor variations had been tried. The matter has never, in fact, been settled completely, and differing opinions are held even today. Nowadays, however, it is only relatively small changes in shape and angles that will be found. In Gibbons's time, although the furnace lines were known to have an effect on furnace operation, there were certain fundamentals that nobody liked seriously to question.

One of these fundamentals was the square shape, in plan, of the furnace hearth and crucible. This shape was repeated in new furnaces and in rebuilds of old ones, without question and it is easy to see how it had originated. In early blast furnaces the hearth proper was made of a single large flat stone and the crucible was built of the largest possible pieces, since it was easier to cut and dress a small number of relatively large blocks than to prepare a large number of small ones. It is also easier to cut a piece of stone in rectangular form than to shape it to form segments of a circle. So the hearths were made square. Later, when refractory bricks became available, shaping the hearth became much simpler, for the bricks could be laid just like ordinary building bricks, which they resembled closely in size and shape; they were, of course, fire-resisting, and were laid in a refractory mortar.

There is some evidence to show that the idea of departing from the traditional square hearth was tried in Shropshire, but with what result is not known. Gibbons, however, made his hearth round instead of square, and immediately found that his furnace worked much better. He was led to carry out his experiments by simple, practical observation. At one time Gibbons had charge of six blast furnaces in Staffordshire ironworks owned by his family business, and he noticed two important facts in the furnaces he controlled. Firstly, a new or relined furnace always ran badly at first, giving a much poorer output than when it was nearing the end of its campaign, all other things—ores, fuel, blowing rate and burdening (the ratio of ore to fuel)—being equal. Secondly, when the furnace was blown out, the hearth, which had been built in the traditional square form, had undergone a rough rounding-off by the action of the fire. Looked at from above the neat square corners had gone and the straight-sided walls had been gouged out until the hearth was roughly circular in form. The inference, to Gibbons, was obvious; the furnace seemed to prefer the rounded hearth. He decided in 1832 to see what happened if he built a hearth of this shape deliberately and his experiment was highly successful.

To get a fair comparison Gibbons built a new furnace, with a circular hearth, alongside an existing one of the traditional pattern. Buth furnaces were blown by the same engine, both were charged with the same materials. After six months of operation Gibbons compared the output figures. The old furnace, with a square hearth, produced 75 tons of iron a week, an average figure for the period. The new furnace, with only one difference, a circular hearth, made 100 tons a week, a figure unknown at the time. Gibbons was encouraged to make further experiments, and he tried a bigger hearth, steeper bosh angles and a higher stack. Again the changes more than justified themselves.

In 1838 another Black Countryman, T. Oakes, took Gibbons's idea still further. Oakes built a furnace with a hearth 8 ft diameter, raised the total height from the customary 40 ft or so of the time to 60 ft, and stepped up the blast pressure from the normal figure of about 1 lb/in^2 to 4 lb/in^2. Oakes, like Gibbons, was highly successful, and his furnace soon broke all records with an output of 236 tons a week.

Neither of these two pioneers in blast furnace operation sought any patent protection; Gibbons, in fact published a small book about his experiments. In the then rapidly developing Black Country iron industry there were many furnace operators who followed suit, and the new idea of the bigger, circular hearth was taken up enthusiastically. As is the case with so many novel developments, there were some ironmasters who would not take the plunge, preferring to stick to what they had always done, and this attitude persisted in some places for quite a few years.

But the round hearth, the bigger furnace, higher blast pressures, and a larger number of tuyeres had such manifest advantages that progressive managements adopted them, and they became the basis of the very large furnaces of today. Hearths other than of circular form, however, were not quite forgotten. A few of rectangular or even oval form appeared in the nineteenth century, and long narrow rectangular hearths are used today for smelting some non-ferrous metals. For iron, though, the circular hearth pioneered by Gibbons and Oakes gradually became the standard for blast furnaces which it remains, everywhere, though it would be unwise to say that it will never change again.

Gibbons brought another innovation to blast furnace operation, too. He introduced a new material into the charge. In many parts of the country where ironmaking had gone on for a long time there were heaps of slag or cinder as it was often called, discarded by users of old and inefficient methods of making wrought iron. Cort's dry puddling process, widely used by the time Gibbons carried out his experiments, was wasteful, and the slag contained quite a lot of iron. So did the slags from the older processes. Gibbons charged some puddling furnace slag or cinder into his furnace along with the usual iron ore, and once again he succeeded in his experiments.

He found that a good proportion for the charge was half or two-thirds cinder with the ore; this gave him a good iron. There was no reason why it should not, though some of his fellow ironmasters held Gibbons to ridicule for they contended that he could never make a really good pig iron by using such a large proportion of 'rubbish'. Of course, the truth of the matter was that the cinder was only rubbish because nobody had hitherto found a use for it. Gibbons benefited too, from the fact that the cinder was cheap; it only had to be dug out of heaps, not mined from underground.

Cinder in the blast furnace charge became quite common in some areas in the nineteenth century, though it never quite got rid of its stigma. Those who used it described their iron as cinder pig, or part-mine iron, while those who used ore alone called theirs all-mine iron. Cinder pig remained common for most of the nineteenth century and only ceased to be of importance when the old cinder heaps were consumed. More efficient methods of wrought iron making, developed in the first half of the nineteenth century, did not produce such large quantities of iron-bearing cinder.

At roughly the same time as Gibbons and others were developing the blast furnace and its charge, a Black Countryman, Joseph Hall (1789–1862) was making some experiments on wrought iron production which were eventually to revolutionize this section of the trade. Hall was led to investigate wrought iron production by his

9 Four 19th century blast furnaces at Cyfarthfa in Wales, charged by an incline between the middle two

10 Two 19th century blast furnaces at Netherton, Dudley, Worcestershire. One, on the left, is a normal masonry furnace; the other is of the iron-cased type

11 A typical pig bed at casting

12 The Earl of Dudley's Round Oak works at Brierley Hill, Staffordshire, a characteristic wrought ironworks of the mid-19th century

observation of the fact that his employer, a small-scale producer of wrought iron by Cort's dry puddling process, was careful to the extent of being miserly in saving and re-using odd scraps of iron, yet apparently quite unaware of the wastefulness of the process itself.

Hall knew that Cort's process needed a lot of pig iron to make a ton of wrought iron—about 40 cwt of it, in fact. The fault lay in the method of decarburization. Cort used atmospheric air as his decarburizing agent, and ignored the other chemical reactions in his furnace. He (and all those who adopted his process) used a cheap and convenient material, sand, to line the working hearth or bowl of the puddling furnace. Sand resisted the heat of the furnace but it had a very bad effect on the process. When the iron in the furnace was brought up to melting point, some of it formed magnetic oxide (Fe_3O_4), which made slag with the silica in the sand. This was a continuing source of waste, and quite a lot of metallic iron was lost. Hall was not the first to observe the wastefulness of the dry puddling process, though he was the first to make a commercial success of the solution.

S. B. Rogers, of Nantyglo, South Wales, suggested as early as 1818 that an iron oxide should be used on the puddling furnace hearth instead of sand. It was a sound idea. As the molten metal was now in contact with an oxide-rich substance, there was plenty of oxygen available for the decarburizing reaction, and the serious wastage of iron into the slag was eliminated. Rogers failed to interest the ironmasters of South Wales in the idea, however, and the records of the time merely serve to confuse, for it was said that Rogers was trying to use an *iron* bottom (in fact he gained the nickname 'Old iron bottom'), whereas he was actually proposing the use of an iron *oxide* bottom—a very different thing. There is evidence that Rogers tried his idea out, but it is certain that it never achieved commercial success.

Hall's first experiments were made in about 1816, that is at about the same time as Rogers's, but there is nothing to suggest that there was ever any contact between the two men, or that Hall had any knowledge of what Rogers was doing. Working on his own, in his employer's works, Hall decided to see what happened if he put a load of what was called bosh slag or bosh cinder in the puddling furnace. He knew that there must be iron in this bosh cinder, which accumulated in the water troughs or boshes used to cool the working tools at the puddling furnace.*

He was right, but the effect of heating the charge was rather more spectacular than he expected. The charge began to boil, then boiled violently, overflowing from the furnace to the floor. Hall could do nothing but watch and hope for the best, and as he did so the boiling ceased, and the contents of the furnace lay quiet. When he looked inside, he found that the furnace contained, as he had expected, what looked like usable iron. He gathered together the spongy pieces, took them to the hammer and hammered, or shingled them into a ball. It was the best wrought iron ball he had ever seen.

Without really knowing why (at first, at any rate) Hall had discovered a vastly improved method of making wrought iron. The bosh cinder he had used certainly contained pieces of iron, but it also contained quite large quantities of iron oxide, which for the furnace reaction, was very important. Here was a readymade decarburizing agent, cheap and easily available. When the charge was heated, the oxygen in the iron oxide reacted rapidly with the carbon in the iron, forming carbon monoxide; this very rapid reaction caused the charge literally to boil. It was this which gave the name pig boiling to Hall's process. This was its commonest name, though it was also called wet puddling, to distinguish it from the older dry puddling invented by Cort.

Pig boiling was adopted both quickly and widely, and it eventually became the only method of wrought iron manufacture in Britain. There was every inducement to use it, for it was faster, more economical of iron, and made a better product. The saving in iron was impressive. Hall's process used about 21 cwt of pig iron to make a ton of wrought, against 40 cwt/ton with the older method. It not only removed carbon effectively, it took out phosphorus as well. This was because there was no build-up of an acidic slag as in the dry puddling process. Acidic slag came from the sand used in the older process. In Hall's process there was no sand, and the slag remained chemically basic. This type of slag combines with phosphorus, which was unwanted in the iron; it will be remembered that it makes iron cold short. Because it was such a good decarburizing process pig

* This is a separate and distinct use of the word bosh. Its meaning at the blast furnace has already been discussed. In the forge (and, for that matter, in a blacksmith's), bosh always meant a cooling water tank or trough.

boiling made it possible to dispense with the refinery, and iron straight from the blast furnace could be used. Besides eliminating a complete process, refining, pig boiling was in itself quicker than the older puddling process, and here again there was a saving.

Hall did run into one difficulty, but he soon overcame it. He found that bosh cinder, while wholly suitable for decarburizing the charge, attacked the cast iron plates forming the bottom of his furnace. So he roasted a quantity of cinder and found that in this form it was much less fusible. A layer of roasted cinder protected the bottom plates very effectively and a layer of bosh cinder on the top of the roasted product reacted with the iron in the charge. Hall never patented his pig boiling process but he did, a few years later, take out a patent for roasted cinder, which he called 'Bull-dog'. The fact that this patent was dated 1838 has led some people to think that the pig boiling process originated at this time, but in fact it was widely used by at the latest the late 1820s. By the middle of the nineteenth century pig boiling was the only process of any real significance used in Britain for making wrought iron, and it remained in use until wrought iron itself finally succumbed to the onslaught of the cheaper mild steel. A few makers of high-grade wrought iron continued to use Cort's process well into the present century, but their total output was small. It should not be assumed that because a small number of wrought iron firms of high reputation carried on using the older process that this process had any special virtue. It had not, and its relative wastefulness put it out of court for most producers. Those who kept on using Cort's process did so principally because the established reputation of their iron could command a higher price than that of their competitors. Manufacturing economy was not therefore of such importance. Trial of a new method of manufacture, on the other hand, seemed to them to be a risky procedure, and they kept on in the old way because they knew it worked.

Minor changes were made to Hall's pig boiling process as time went on, but these were mainly confined to the materials used for decarburizing. These materials—the fettling of the furnace—could really be any one of several which were easily available; good quality iron ores, for example, or mill scale (the pieces of oxide which flaked off wrought iron being forged or rolled) could serve just as well as the older 'bull-dog', which fell out of use in due course.

Concurrently with Hall's work in the wrought iron making section of the trade there were developments in the finishing departments, and some further changes in operating blast furnaces. The latter were to succeed only at a later date (by about the middle of the century) and will be considered in more detail at the appropriate point in the story, but some of the other developments are best dealt with here.

One of the factors in ironmaking which began to receive attention in the first quarter of the nineteenth century was fuel economy. It had not aroused much interest previously and it only gained ground slowly, since the ironworks were now mostly in areas where coal was cheap and there was little real inducement to try to use less fuel. Economy in the use of fuel was taken very seriously in some other industries (in Cornish mining, for example, for there was no coal in Cornwall), and it came to be recognized as important in the iron trade in due course. Today it is vital. But its beginnings were small.

A lot of heat is used in making and working iron, and in the early years of the nineteenth century a few people realized that much of it could be used more efficiently. A puddling furnace, for example, could waste a lot of heat up the chimney stack, and so could a mill furnace, which was similar in design, but larger, and was used for reheating iron for rerolling. John Urpeth Rastrick (1780–1856), devised a very good way of making use of some of this waste heat. Rastrick, who is better known as a railway engineer, did some good work in the iron trade, and in 1827 introduced his waste-heat boiler. This was of a very specialized design, particularly suited to the puddling furnace, and was usually used in conjunction with two or four furnaces. The Rastrick design of boiler found little or no application outside the iron trade, where it was employed very extensively, but the significance of Rastrick's invention is much greater than the restricted application of his particular design of boiler suggests. To him must go the credit for the idea of using the waste heat of an iron furnace to generate steam, though waste-heat boilers had been tried earlier in other fields. Waste heat recovery could be, and was, applied in many other industries besides ironmaking, and many other forms of boiler were used. Waste heat steam generation, incidentally, is still of importance, though other means of recovering waste heat have assumed more importance since the steam engine has given way to other prime movers. In Rastrick's time the whole of the ironworks machinery was steam driven, and there was always a ready use for steam. His boiler, specially suited as it was to ironworks furnaces, was a good one; it lasted in quite large numbers in ironworks well into living memory, and a very small number may yet be found in use.

Better means of producing wrought iron and more economical operation of the furnaces, coupled with an ever-increasing demand for iron meant, obviously, that more iron was made. But the increasing demand brought with it something else which was to have a profound effect on the industry—a need to meet higher standards in the products themselves, as well as to make more of them. This again is something which has continued to the present day and goes on unabated. Quality standards are still rising.

Two developments in the iron rolling mills in the first twenty-five years of the nineteenth century stemmed directly from the increasing demand for production and for better product quality. Both were to become of great importance; both are the basis of production processes used widely today. One was the three-high rolling mill, the other the guide mill.

Rolling of iron was done, as we have seen, between two rolls mounted in frames or housings. Such a mill, to distinguish it from later types, particularly the one which we now have to consider, will be known henceforward by its proper name—a two-high mill. The two-high mill is simple, it can be made robust and can do a lot of good work. It has its limitations, however. To see what those are, it is necessary to consider how a piece of iron (technically, the piece) is rolled. In rolling the simplest shape of all, a flat strip, in the simplest possible two-high mill, a single stand with one pair of rolls, the piece would be passed between the rolls, returned to the starting point while the rolls were set a little closer together, and then rolled again. This would continue, the rolls being screwed down closer every time, until the required thickness was obtained. To roll a round or square, the piece would be passed through pairs of half-round or half-square grooves cut in the rolls and matched in position in such a way that they formed round or square gaps between the rolls, each one a little smaller than the last. So the piece would be rolled in one groove, returned to the starting point, rolled in the next groove, and so on until the required size was obtained.

Each passage through the rolls, when work was actually done on the piece, was called a live pass; each return movement was a dead pass. These terms are still current. It will be seen that during about half the time the piece spent at the mill, no work was done on it at all; it was merely coming back in a dead pass to start the next live one. This, though wasteful of time, was acceptable when the product being rolled was of reasonable size, say 2 inches in diameter or more and not more than a few feet long. If the product were small, say a $\frac{1}{2}$ inch diameter bar, it cooled too quickly, as a result of all the dead passes, to be rolled properly to any reasonable length. Larger pieces could be reheated at some time during the rolling sequence; smaller ones could not, for the smaller the cross-sectional area the greater the length, and they would be too unwieldy to deal with in the furnaces of the time.

The problem of rolling smaller sections was solved by the three-high mill, which is essentially simple and very effective. It is not known exactly when it was introduced or by whom, but the evidence points to its origin in the Black Country area of Staffordshire at some time prior to 1815. Several such mills were known to exist there at this date, a fact which disproves the attribution, in some sources, of the mill to John Fritz, an American, in 1857.

In the three-high mill a third roll, of the same size as the other two, was mounted over them in the same housings, and the grooves, or passes, in the rolls were so cut that the piece passed first between the bottom and middle rolls, and came back between the top and middle rolls. Work was thus done on the piece in both directions, there were no dead passes, and the time of rolling was halved. This saved heat, and the piece could be rolled out smaller in cross-section and longer, before it cooled too much for rolling.

The second development in iron rolling, the guide mill, solved another problem, that of misshaping or finning. It might be supposed that if the grooves or passes in a pair of rolls were cut to the exact size required, and the hot iron passed through them, it must emerge also to the exact size required. This is neither true in theory nor in fact. It could only be so if the various components of the mill were infinitely stiff. When a piece of iron is rolled it is squeezed, and this squeezing produces an opposite reaction on the rolling mill. The housings stretch and the rolls bend. These movements are very small, but they are sufficient to allow the iron to squeeze out slightly at the point where the roll passes match, and form small fins on opposite sides of the rolled product. Fins, however small, were unacceptable to some users of rolled iron. They could be eliminated in the larger sized rolled products by rolling the piece several times through the final, or finishing pass in the rolls, turning it through 90° between each pass. This could not be done with small sections, which would cool too quickly.

The answer to this problem was to roll the product, if it were to be a round section, into an oval of the same cross-sectional area in the last pass but one. In the final pass the change in the product was one of shape only, and not of cross-sectional area; there was therefore no metal available to form fins. If the required finished section were a square, the penultimate pass was designed to produce a diamond of the same cross-sectional area. Again there was no finning when the finishing pass changed the shape to a square.

To carry out the finishing pass properly it was necessary for the piece to be entered into the rolls with the long axis vertical, and it was so guided by a suitably-shaped hole in a metal box, fixed in front of the rolls. This, the box guide, or simply the guide, gave the names guide mill to the mill and guide rolling to the process. Guide rolling is still practised extensively and the oval-to-round, or diamond-to-square rolling sequence is also used for metallurgical reasons at various points in the overall sequence of rolling of some products. It might be noted in passing that many other mills (in fact most of them) have metal devices called guides for ensuring that the piece is correctly presented to the required pass, but the term guide mill is restricted to one which uses the oval-to-round or diamond-to-square sequence to avoid finning.

There is no record of when the guide mill first came into use, or who invented it, but, like the three-high, it almost certainly originated in Staffordshire before 1820.

The events so far chronicled in the wrought iron side of the iron trade had set the scene for the pre-eminent position wrought iron was to occupy in the middle of the nineteenth century, and it will be necessary to consider the trade in some detail. Before doing so, however, to preserve the chronological sequence, mention must be made here of a further development, already referred to in passing, at the blast furnace. This was the use of waste gas.

It had not escaped observation that the gas issuing from the throat of the blast furnace was potentially useful. It burnt at the throat with a long luminous flame that attracted the attention of visitors to the ironmaking districts. In 1834 an attempt was made to take off some of the gas at a Wednesbury, Staffordshire, blast furnace, and put it to some use, but it failed, and it was not until 1845 that some modest success was achieved. In that year J. P. Budd, of Swansea, put a hot-blast stove at the top of one of his furnaces, and led the waste gas through it, so heating the air for the furnace. It worked, but was only moderately successful, and although Budd was granted a patent in 1845, little more was heard of his idea.

Others tried to collect the gas, or some of it, lead it down to ground level, and burn it in hot-blast stoves or under boilers. This was a logical thing to do (and it is done today) but there were many problems in the actual gas collection. No real success is recorded until George Parry, of Ebbw Vale, Monmouthshire, devised a means of closing the top of the furnace in 1849. His device was the bell and cone, which are used in modified form today.

Parry fixed an inverted cone, or hopper, in the throat of the furnace and arranged inside it a cast-iron bell, which could be raised and lowered on a counterbalanced arm. When the bell was raised, it made contact with the underside of the hopper and sealed the throat of the furnace. All the gas was thus compelled to go into a large pipe (the offtake) built into the top of the furnace, and thence, via a vertical pipe (the downcomer) down to ground level. Here it was burnt in hot-blast stoves and boilers. Furnace charges were dropped on to the bell, and when sufficient had accumulated, the bell was lowered, allowing the charge to fall into the furnace.

Thus for most of the time the furnace top was completely closed, and the whole of the gas output was usable. Only at intervals of say, half an hour, was the gas allowed to escape momentarily when the bell was lowered. A small coal fire was kept burning all the time under the boilers or stoves to ensure that the gas was re-ignited after the bell had been lowered, but apart from this no fuel was needed. Like all new ideas the bell and hopper was not accepted universally at first, and it gave trouble in the early days on some furnaces. But it was far too obvious an improvement to furnace operation to escape the notice of the progressive ironmasters, and it spread quite rapidly.

CHAPTER FIVE

Wrought Iron Supreme, c. 1860

Wrought iron was in its heyday in about the middle of the nineteenth century. Its decline thereafter produced no change of significance to report, other than a steady reduction of output, from that time until the present day, when wrought iron is effectively extinct in Britain and all other heavily industrialized countries. Production methods went on unchanged and because of this it is possible to record, for the first time in the history of ironmaking, the technicalities of a branch of the trade which are not only known from written records, but can be supplemented from the memories of people still living. With minor exceptions in the matter of plant details, wrought iron making was the same in 1960 as it had been in 1860; very much less was being made, that is all. How great this fall in output had been can be seen from the records of the Black Country, formerly the most important wrought iron producer in Britain.

In 1865 the Black Country had 2,116 puddling furnaces (this figure actually rose a little to reach its maximum of 2,155 in 1872). At the beginning of 1960 there were only four or five puddling furnaces in the area and in March of that year those remaining few closed down, leaving the Black Country without a single wrought iron producer. A few puddling furnaces remained in operation elsewhere for a few more years but at the time of writing (1967) the number left in the whole country could be counted on the fingers of one hand.

Wrought iron in its heyday was available in more than one grade and in a surprising variety of shapes and sizes. All these, since they are now little more than a memory, are worth considering in some detail, as is the puddling process itself.

There was never, in the matter of size, a typical wrought ironworks. Some had as few as four or five puddling furnaces and a single rolling mill. Others had several mills, served by twenty to fifty puddling furnaces. Barrows and Hall, at Tipton, Staffordshire (the firm founded by Joseph Hall) had 100 puddling furnaces in three works. The Consett Iron Company, in Co. Durham, had over 100 furnaces in one works. But if there was no uniformity in size of works, there was a more or less standard proportion of puddling furnaces to rolling mills; five or six to a mill was the normal figure.

A puddling furnace had a crew of two, the puddler, who was in charge, and the underhand, who was usually learning the trade. The puddler was responsible for the whole operation of the furnace and for seeing that the iron was of the right quality. Production of bad iron (which was easy if proper skill and care were not exercised) was punishable by fines and, of course, if it occurred too often, by dismissal. In the nineteenth century the men worked a shift or turn of twelve hours, usually from 6.0 a.m. to 6.0 p.m., or from 6.0 p.m. to 6.0 a.m., Saturday afternoons and Sundays being free. In this, of course, the men were only in line with shift workers in other industries. By the time the trade was virtually at an end, the turns were of eight hours, the shift-change times being 6.0 a.m., 2.0 p.m., and 10.0 p.m. Often, owing to shortage of men or orders, only a single turn was worked in the day.

Working a complete charge, from pig iron to wrought iron, in the puddling furnace was known as a heat; it lasted about two hours, so a puddler could normally get six heats in a twelve-hour shift. A heat started with the puddler attending to the lining or fettling of the furnace bowl, to prepare it to receive a charge of pig iron. How much he had to do at this stage depended on circumstances.

If the furnace was newly repaired or newly built (work which was done by a specialized bricklayer) the puddler would have to fettle completely, or, as he said, make a bottom. To do this he first of all got the furnace to full heat, by opening the damper in the flue leading to the stack or waste-heat boiler. He then threw in a layer of bosh cinder or rubbish from the troughs used to cool the working tools, and followed this with a second layer of fettling, this time of broken cinder (i.e. puddling furnace slag), together with some mill scale, and perhaps a little iron ore.

Fettling completed, the puddler and underhand threw in a bundle of wrought iron scrap (cropends, or ends cut off rolled iron products to square them up were typical). This scrap was worked up into a finished ball, at welding heat, of wrought iron, and the effect of this was to settle and consolidate the fettling or, as the puddler said, to glaze it. The ball was then taken out, and the normal heat could commence. If the puddler had just worked a heat, preparation for the next one was much simpler. All he had to do was to throw in a few shovelfuls of light fettling, such as mill scale.

When the fettling was in order, the furnace was charged, the underhand throwing in, one at a time, about 5 cwt of cold iron pigs. The working door was then closed, and melting-down commenced. This lasted about thirty minutes, the underhand turning the pigs over every few minutes, by means of an iron bar (the paddle) inserted through a small hole at the bottom of the working door; this promoted melting.

As soon as the iron was melted it began to react or 'work', with the fettling. Decarburization had begun, and it was necessary to prevent over-oxidation, which would only burn and damage the iron. A non-oxidizing or reducing atmosphere was therefore created in the furnace, by lowering the damper and allowing the furnace chamber to fill with smoky flame. This period, which lasted about ten minutes, was known as smothering, and was the beginning of the critical part of the process. Too much heat or air, from now on, could spoil the iron, and the skill of the puddler was the only safeguard. He watched his iron carefully, using the damper often to control the heat and atmosphere.

When smothering had been in progress for about 10 minutes the working, or reaction, with the oxide fettling became more pronounced. Now, since the iron was losing its carbon, the melting point was rising, and the damper was raised a little, to give more heat in the furnace, but not enough to remove the reducing atmosphere. Thus the most interesting part of the puddling process, the boil, commenced. Reaction was now quite violent, and the bubbles of carbon monoxide gas (CO) broke through the iron, burning at the surface to carbon dioxide (CO_2) in countless little jets of blue flame, known as puddlers' candles.

For the whole of the boil the iron had to be kept in motion, the puddler and underhand taking turns of a minute or two, stirring or puddling with a hook-ended iron bar called the rabble; the stirring itself was also called rabbling. The boil lasted about thirty minutes, and, since the iron in the furnace became progressively stiffer and more pasty as it lost its carbon, it was the hardest part, physically, of the whole process.

There was no mistaking when the boil had finished, for the puddlers' candles ceased suddenly; almost dramatically. No more carbon was left in the iron, which, the puddler said, using a term current since Cort's time, had come to nature, or come to grain, and it was necessary to take it out of the furnace as quickly as possible.

So the puddler took the paddle (the iron bar also used during melting down) and quartered the pasty iron, or divided it into four pieces roughly equal in size. Then, one by one, the pieces were gathered, or balled up, into balls, each weighing in the region of 100 to 112 lb, and taken out, with the aid of large tongs. There was usually a fifth ball, rather smaller than the others, made up of the scraps of iron which remained on the furnace, and this was taken out in the same way, so leaving the furnace empty except for a quantity of molten slag or cinder. This was tapped out through a tap hole below the working door, and the furnace was ready for fettling and working the next heat.

The pig iron charged to the furnace contained, in addition to carbon; manganese, silicon and phosphorus. All these were removed in the puddling process, a summary of which is as follows: In the first, melting down period, most of the silicon and manganese and some of the phosphorus were oxidized and passed into the cinder. During the second period, smothering, the remaining silicon and manganese and some more of the phosphorus were taken out. Too much heat and oxygen at this stage might have eliminated the carbon too soon, leaving phosphorus in the finished iron; hence the reducing atmosphere, or smothering. After the smothering, the boil, or third period, removed the carbon and the remaining phosphorus. The actual working period was about 100 minutes per heat, which allowed a little time at the beginning and end of the heat for fettling, slagging and cleaning up, so a heat was completed comfortably in the normal two hours.

The balls as they left the puddling furnace were wrought iron, but they were of no commercial use in that form, being a shapeless lump of iron mixed with a lot of molten cinder. They were therefore taken immediately to the hammer, and there shingled, or hammered to consolidate and weld together the particles of iron, to expel surplus cinder, and to shape the ball into a rectangular bloom, in which form it was suitable for rolling.

WROUGHT IRON SUPREME, c. 1860

At the time of which we are writing, the middle of the nineteenth century, the standard equipment for shingling was the helve hammer. This, as has already been pointed out, should not be confused with the earlier tilt hammer. It was heavier, moved more slowly, and was made of cast iron, with wrought iron working faces dovetailed in. A helve had the fulcrum at one end, the other end being lifted and allowed to fall by cams on a cam barrel turned by a steam engine. In between the fulcrum and the cam barrel was the hammer face, which was directly over a fixed anvil. There was no spring device as in the tilt, the hammering was done by gravity. A typical helve used for shingling weighed from 5 to 8 tons, so heavy blows were given. The speed of rotation of the cam barrel was so arranged that the helve gave one blow a second, thus giving time for the shingler to manipulate the ball between blows.

Because the cams of the normal ironworks helve lifted it at its outer extremity it was known as the nose helve (or frontal helve), which distinguished it from a variant, the belly helve. This had the cam barrel between the working face and the fulcrum. It found little use for shingling in the ironworks, but was favoured by the makers of wrought iron forgings, because, as the working face was at the extreme end, a forging of relatively complicated shape could be moved around easily between the hammer and anvil.

Both types of helve were usually driven by a steam engine (occasionally a waterwheel) which drove other machinery at the same time, and in neither case was the prime mover stopped to stop the helve. Stopping, which was often necessary, as balls did not come continuously from the puddling furnace, was achieved by a process known as gagging-up. To gag up the helve, the shingler placed an iron bar on one of the cams as it approached the hammer. This caused the cam to lift the hammer higher than it would normally go, and at the top of its lift the shingler swung into position a pivoted iron prop or gag. The cam barrel then rotated freely and the hammer remained gagged-up until it was wanted again. To set it in motion the bar was again placed on a cam and the hammer was thus lifted off the gag, which was dropped down out of the way.

Such was the normal apparatus used for shingling in the heyday of wrought iron. However, it had a competitor, the steam hammer, which, by the 1860s, was beginning to be of importance. The steam hammer had been invented by James Nasmyth (1808–90) in 1839, but was only adopted slowly in the iron trade because, at first, there was considerable opposition to it. This, in some ways, was not unreasonable. The force of the blows of a helve cannot be controlled, but those of a steam hammer are easily variable. Opponents of the steam hammer said that it could be used, by collusion between the puddlers and the shingler, to 'nurse' bad iron with gentle blows, and so get it away from the forge and into the rolling mills. It is true that sooner or later the bad iron would be discovered, but by that time it would be difficult to put the blame on any particular puddler. With the helve, the heavy, unvarying blow would disclose immediately whether a ball was good or bad; a bad one would crumble. All this had some truth in it, but it was equally true that good management would also ensure that good iron was made, and the steam hammer was so convenient to use that it made its way into the ironworks in spite of its detractors. In due course it virtually superseded the helve, though a few of the latter lingered on until the 1920s. Shingling expelled large quantities of molten cinder, and to protect himself the shingler wore sheet iron foot and shin guards, a heavy leather apron, and a wire gauze face mask.

Wrought iron balls hammered or shingled into blooms had next to undergo rolling, and they passed, while still hot enough to roll, straight to the first of the rolling mills, the forge train. This was a simple, two-high mill, with rolls usually about 18 or 20 inch nominal diameter.* At the forge train the short rectangular bloom was rolled down, in a series of passes to a flat bar about $\frac{3}{4}$ in or 1 in thick, 4 in to 6 in wide and 12 ft to 15 ft long. It was then allowed to cool, for this first product of the rolling mill, puddled bar, or muck bar, was still not of commercial value, and it was not sold in this form. Muck bar was rough and of poor mechanical strength and could only be counted as the raw material for further processing.

Wrought iron shares with some other metals the interesting property of improving physically as mechanical work is done on it, and this was the basis of the grades of iron produced for sale. There is a limit to the improvement gained. Reworking up to about six times gives an improvement in physical properties and after this the

* Roll diameters are nominal because as the roll wears and is re-turned in the lathe to restore the worn profile, it naturally gets smaller. Provision is made in the design of the mill to allow for several re-turnings before the roll has to be scrapped.

quality of the metal falls off again quite rapidly. Actually, reworking was not carried as far as this, for the rate of improvement was too small in the later stages to justify the cost. In practice, reworking was not taken beyond four times, and much of the iron made never got this far. But reworkings were the basis of quality improvement and the market gradings were based on the number of reworkings.

Muck bar, after cooling, was cut up by mechanical shears into pieces about 1 ft or 2 ft long, stacked into what was called a pile, reheated in a mill furnace (which was similar to a puddling furnace in shape, but usually larger), and rerolled. This first rerolling produced the lowest grade or common iron, which was also called crown, or merchant iron. It got the latter name from the fact that it was a good general-purpose iron, stocked by merchants everywhere. If merchant iron was cut up, piled, reheated and rerolled, the product was known as best iron. Reworking of best iron produced best best or BB iron, and the highest grade of all, best best best (or treble best or BBB) iron, was made by reworking BB. This was not common, only a few of the largest and best-known firms made it. Many firms did not go beyond best or BB and some made nothing but crown or merchant iron; this after all, was excellent for many purposes.

In the first rolling, of muck bar, the product was always a flat bar. This was convenient to roll and an easy shape to pile after shearing to short lengths. The rollings which followed were of many shapes, depending on what the iron was to be used for or what further processing, if any, was to follow. Thus, if the higher grades were being made, bars would be rolled at each stage until the last one, when the shape would be that required for the market. If the grade was merchant iron, the required finished shape would be rolled straight from the muck bar pile.

How the pile itself was made also depended on the use to which the finished iron was to be put. Rolled iron is strongest in the direction of rolling, since the fibres are distributed longitudinally. If the finished iron was wanted for making products in which the stress would be mainly longitudinal (chains, for example) the piles were made up by laying the pieces of bar with their axes parallel. This was a plate pile. It was often varied by placing a piece of iron vertically at each side, to box it in, and it was then called a box pile. If the finished iron was required in some form, such as plates for making boilers, where the stress could come in any direction, the pile was often made with alternate layers of bar at right angles. Thus the fibres were distributed to give, as far as possible, equal strength in all directions. This type of pile was called a cross pile, and it was much less common than the other two. Some manufacturers, in fact, were opposed to it, contending that they could make good plates from box piles, but the cross pile was nevertheless to be found in many ironworks all over the country.

No list of the shapes in which finished iron could be rolled has ever been compiled, and it is unlikely that one ever will, for the shapes were legion. They included many which were rolled for special purposes, some of which have long been forgotten. But there were many rolled sections, including some of the special ones, which were in general or at least wide, use, and some of them have survived into the steel age. The specials can be disposed of fairly quickly. They included some with names which are clearly indicative of their purpose, such as rails, hoops, horseshoe, cart-tyre, boot-tip, bucket-handle, can-rim, and hurdle sections. Others were not so obvious: these included star iron (shaped like a star and used in railings); thimble iron, used for rope fittings; key iron, used for engineers' keys, not for locks; and spoke iron, used for building up railway wagon wheels. Many of the special sections were rolled for purposes peculiar to a particular customer or group of customers; in some cases their shapes are known but their functions, now, are not.

One group of special sections was known somewhat indefinitely as 'fancy' iron. Some manufacturers included such things as horseshoe iron in this group, but properly fancy iron meant small bars, flats or other shapes with geometric or other patterns rolled on the surface, purely for decorative purposes. They were used for the manufacture of indoor or outdoor furniture, garden embellishments such as pergolas, and for any other article where a patterned surface, so beloved of the Victorians, was considered to enhance the appearance. Fancy iron is rare today and virtually forgotten, but it was quite common in the period under review, though never made in really large quantities.*

---

* A small collection in the author's possession has patterns including Greek key, overlapping diamonds, square pyramids and a design based on a formalized feather pattern. All were made by rolling in a pair of rolls, one of which had the pattern cut in its surface by hammer and chisel.

13 Water-driven tilt hammer at Wortley Ironworks, near Sheffield

14 Taking the ball out of a puddling furnace

15 Steam engine-driven helve hammer used for shingling. Formerly at Stourbridge Old Forge, Worcestershire

16 Shingling under a steam hammer

The real bread-and-butter lines of the ironworks were those sections such as rounds, squares, flats, angles, tees, strips, plates and sheets which were the raw material of other manufacturing trades. All these have continued into the steel age. But there is one product which is common today in steel which never figured largely in iron. This is the joist or H-section. The joist was first rolled in France in 1849–50, and its use spread to Britain quite quickly, but it was never made in large quantities in iron simply because it was not practicable to do so.

It will be remembered that a wrought iron ball from the puddling furnace weighed in the region of 1 cwt. The rolled bloom would be of about the same weight. Not much of a joist could be rolled from such a bloom, for even a small joist, say 8 in by 6 in, weighs about 35 lb/ft. It is true that two or more blooms could be welded together to make a larger piece, but if this were done the resulting piece would become too large for manual handling. The joist as we know it today is really a product of the steel age. It had to wait until large pieces of metal were readily available, and mechanical devices were designed to manipulate them. Both these developments are chronologically out of place here, and will be dealt with later.

All the sections so far named, with the exception of plates and sheets, were rolled in more or less the same way. After the forge train, which was a two-high mill, the rolling was done in either two-high mills or in three-high, according to the product; usually the larger sections were rolled in the two-high and the smaller in the three-high. Small rounds and squares were rolled, as has already been mentioned, in guide mills. All the movement of the pieces to and from the rolls was done manually; there was as yet no mechanization of handling.

Sheets and plates were in a category all their own. First, it is necessary to define sheets and plates. Today this is quite simple; a sheet is a wide, flat-rolled product up to and including 3 mm (0·118 in) thick. Plates are similar products over 0·118 in thick. In the nineteenth century it was not so easy, for some people used 0·238 in as the upper limit of sheets, and others used 0·125 in ($\frac{1}{8}$ in). Others, again, referred to products rolled in fractional sizes as plates and those in gauge or decimal sizes as sheets. Of all these definitions the last is by far the worst, for 'gauge' might mean almost anything. There were several gauges in use in the nineteenth century and gauges were often to be found which purported to be the same but were in fact different. On the whole the dividing line of $\frac{1}{8}$ in (above this equals a plate) is the best. It was widely used and is reasonably near to the one used today.

Plates and sheets were rolled in the same way and sometimes in the same mills, though it was more usual to keep the products separate. Heavy plates, 4 in or 5 in or so thick, such as those made for armour-plating of warships, were always rolled in mills kept specially for the purpose. Such plates, however, were uncommon, and their production was confined to a few firms, since they called for special techniques and were prodigal of manpower.

A peculiarity of plate and sheet rolling is the operation known as broadsiding. When a product is rolled in grooved rolls it must follow the shape of the grooves, but when the rolls are plain cylinders a different condition applies. Between plain rolls the piece elongates much more readily than it spreads laterally. So if a piece of flat iron were to be rolled straight out it would be extended to the required finished length long before it had spread to the required width. Because of this it was necessary (and it still is, even in modern mechanized mills) to give the piece a certain number of passes first at right angles to its length, to get its width right before rolling it to its finished length. This was known as broadsiding or broadside rolling.

Following broadside rolling, a plate could be finished to the required size by the appropriate number of passes. Sheets, however, introduced a complication of their own. As the sheet got thinner, it naturally got larger in area, and as it was exposed to the air, it cooled more rapidly than was desirable. Moreover, it could get too big for comfortable handling. It was therefore the custom to give a sheet a number of passes and then to double it back on itself and do the remainder of the rolling in this folded or doubled-back form. If the sheet were to be very thin, a further doubling became necessary in due course. When the rolling to thickness was finished the folded edges were sheared off, and the group of sheets (or pack) was separated. The individual sheets tended to stick together to some extent, and a long blunt knife or opener was used to separate them. As the separating generally bent the sheets, one or two flattening passes were given in the rolls to smooth them out.

Because sheets were rolled, in the thinner sizes, by doubling once or twice, they were given names based on this operation. Sheets rolled without doubling (the thicker ones) were singles, those which were doubled once, doubles, and those which had a second doubling, triples or lattins. There has been much speculation, incidentally, about the alternative name, lattin, but so far no satisfactory explanation for it has emerged.

Dimensionally, singles were usually from 3 Gauge to 20 Gauge (0·252 in to 0·036 in), doubles were from 21 G to 24 G (0·032 in to 0·022 in), and triples were from 25 G to 27 G (0·020 in to 0·0164 in). Occasionally sheets were made thinner than 27 G; in such cases they were usually called double doubles. It will be noted from the sizes given above that in the singles range the thickest sheet was really what we should today call a light plate (it was over $\frac{1}{8}$ in or 3 mm), but as has been said, the dividing line was arbitrary and variable.

Sheets provided an example of the rather rare practice of cold rolling of iron. This applied particularly when they were required to take a good surface finish, as for tinplate, enamelling, or japanning and decorating. Separate rolls, with very smooth, polished surfaces, were kept for cold rolling. Sheets which had had cold work done on them became hard and brittle, and it was necessary to soften them by annealing. This was done in an iron vessel, to exclude as much air as possible, and so prevent oxidation of the surfaces of the sheets. Usually, and always for tinplate making, the sheets were then given a short dip in sulphuric acid to remove such surface scale as had built up. This gave rise to trade terms, current until quite recent years as CRCA (Cold Rolled, Close Annealed) sheet, or CRCA P & O (the same plus Pickled and Oiled, the oil being for temporary protection).

Such were some typical products of the heyday of wrought iron (and, for that matter, for the rest of its existence). It is now only necessary to say something of the actual rolling machinery itself; this will help to set the scene for what followed in the next few years.

In construction the two types of rolling mill then in use were similar. A mill consisted of one or more stands, each stand comprising a pair of heavy cast-iron housings or frames, two (or three) rolls, and the necessary bearings and means of driving the rolls. Each roll had a working surface or barrel, on which the passes were cut (or which was left plain if for sheets or strip), and was reduced in diameter at the ends to form necks, or bearing surfaces which ran in the bearings. At the extreme ends of the rolls were cruciform shapes or wobblers, which engaged with hollow cruciform couplings or wobbler boxes; these transmitted the drive, via spindles and a gearbox or pinion box, from the engine. The roll necks ran in bearings in sliding blocks called chocks in the housings, and screws enabled the chocks to be positioned in the housings as required. The wobbler boxes working loosely on the wobblers and spindles acted as rough and ready universal joints, and allowed the rolls to be raised and lowered in a vertical line relative to each other while the centres of the pinions in the drive gearbox remained fixed.

It was thus possible to turn off the worn faces of the rolls in a lathe and use them again in the mill, or in the case of a sheet or plate mill, to separate the rolls relatively widely to allow a thick piece to enter for the first pass and then screw them closer together progressively as the piece was rolled.

A pair of housings complete with rolls, chocks, bearings and screws was called a stand, and a number of stands went together to form a mill. How many stands made up a mill depended on circumstances. For some purposes a single stand might suffice. This could be so in a forge train, where a single pair of rolls could contain sufficient passes to reduce a shingled bloom to muck bar. The same could apply with the two-high mills used for finishing, but in both cases it was often found that there were actually two or three stands, connected together end to end, and driven by the same engine. This made it possible to keep a variety of passes in readiness, and so to roll various sizes or shapes of product without having to change rolls.

For light bars and sections, where the finished product was small in cross section a relatively large number of passes would be needed, and the usual equipment in these cases was three or four stands, the first two or three being three-high and the last one two-high, with a guide pass for finishing to final form. Sheet mills usually comprised two two-high stands, one for roughing or rolling the first few passes and the other for finishing. Cold mills for sheets were generally kept quite separate from the hot mills and were often in a different part of the works altogether.

The mills were driven, of course, by steam engines, and a common arrangement was to have a rotative beam engine driving the forge train (18 in to 20 in rolls) on one side of the engine house and a two-high bar mill of similar size on the other side of the house. A simple form of dog-clutch was incorporated in the drive to each mill, so that either could be taken out of use if required (for roll-changing, for example) while the other continued at work. In spite of the fact that the high-speed non-condensing steam engine was well established by the middle of the nineteenth century, very little use was made of it in ironworks, where the slow-running condensing beam engine, rotative for the mills and simple reciprocating for blast furnace blowing, held sway.

This was not because of conservatism on the part of the trade. Most of the machinery in the mills ran, of necessity to suit the characteristics of the iron, at quite low speeds. Thus, the helve, as already mentioned, called for a cam-barrel speed of no more than 30 rev/min. Forge trains and two-high bar mills ran at about the same speed, and even the lighter mills for guide-rolled iron did no more than about 300 rev/min.

There was little in the way of auxiliary machinery. Shears were essential, but these, too, were slow running. They were generally of the so-called crocodile type, which can be likened to a very large pair of scissors, and they were driven, usually, by a cam fixed on any conveniently placed slow-running shaft or by a crank and connecting rod with the same power source. Sometimes the shears were driven by a small subsidiary beam coupled to the back of the main engine beam. The total amount of mechanical power used in an ironworks in the mid-nineteenth century was actually small—very small by presentday standards. Mechanical power had to be used for driving the mills and blowing the furnaces; there was no choice. Shears, too, could not be powered by man. But whatever the muscular power of man could do, it was made to do. Labour was still cheap and the individual weights of the pieces to be handled were still relatively small. All this was soon to show signs of change.

However, there was one device which was beginning to find increasing use by the middle of the century which did work at a fairly high speed and for which the slow-speed beam engine was not a suitable source of power. This was the hot saw. First introduced in the Black Country some time before 1840, the hot saw was a form of circular saw which, running at a high speed, trimmed the ends of bars and sections and cut the sections themselves into the lengths required for the market, as they came from the mill, and while they were still hot.

The hot saw was driven by a small high-speed steam engine or occasionally a rotary turbine of the type first described by Hero of Alexandria in the first century A.D. This latter was most wasteful of steam, but in a works where ample waste-heat-generated steam was available this factor was unimportant. The turbine was very simple and easy to maintain, which made it attractive to the ironworks managements.

Hot saws were by no means universal in the mid-nineteenth century and they never superseded the shear, which is still used today. Where they did score was in the clean square cut they made. A shear is bound to squeeze and deform, to some extent, the ends of the bar it cuts. This did not matter when the bars were to be rerolled, as in the case of muck bars for piling, or when the bar was long and thin and the damaged end was a very small proportion of the whole, as in guide-rolled wire rods. When the bar or section was shorter and heavier, however, the damaged end could represent quite a large proportion of the whole. Again, the end might have to be square, as in the case of a railway rail. In such cases the hot saw was ideal.

CHAPTER SIX

The Challenge of Steel 1860 - 1890

The first challenge to the supremacy of wrought iron had come in 1856 when Henry Bessemer (1812–98)* read his famous paper 'The manufacture of malleable iron and steel without fuel' to the British Association. Bessemer's process as he described it was decidedly novel and it caused quite a stir, particularly in scientific circles. But although it is now possible to look back on the events of the time and recognize the year 1856 as marking the beginning of the steel age, the new process did not seem at first to be a very serious competitor to the established order of things in ironmaking. This was particularly so because after a spectacular debut the Bessemer process ran into trouble and failed for a time to live up to its promise. Why this was so we shall see in due course, but before looking at the Bessemer process in detail it is necessary to consider briefly what was meant at the time by 'malleable iron and steel'.

Malleable iron can be dismissed quickly, for in using the term in the sense he did Bessemer was following a bad practice of the time and misusing the word. By 'malleable' he meant wrought iron. Malleable iron, correctly, was cast iron which had been subjected to heat treatment to render it less brittle, and the annealing process by which iron castings were made malleable was invented by the Frenchman Réaumur in 1722. It is still used quite extensively today and the term malleable is now confined to cast iron of this type. In the nineteenth century the term was often applied loosely to wrought iron. So much for what Bessemer meant by malleable iron.

Steel was a different proposition. It was no novelty at the time, though 'steel' then meant carbon steel. What Bessemer did, in fact, was to invent a new process which produced a new material—mild steel. Carbon steel was made at the time of Bessemer's invention by

* He was not knighted until 1879.

a method then more than a century old. This was the crucible method of Benjamin Huntsman (1704–76). Huntsman, a clockmaker, of Doncaster, was dissatisfied with the quality of the only steel he could get and set out to improve it. The steel available to Huntsman was made by the cementation process, the history of which is not well documented, but it is known to have been in use in 1722 and probably much earlier.

Cementation was carried out by packing bars of the purest available wrought iron with charcoal in closed clay vessels and heating the pack in a furnace for several days. Carbon from the charcoal diffused slowly into the wrought iron, so giving it a carbon steel skin, which could be hardened. Because the bars on being taken from the furnace had a blistered surface the product was known as blister steel. This was sometimes used as it was. Sometimes it was broken into pieces, piled, reheated and welded together, so giving a better distribution of the carbon throughout the mass. Blister steel so treated was used extensively for making textile shear blades, from which it got the name shear steel. Occasionally, shear steel was piled and reworked, to make double shear steel, of, theoretically, higher quality.

Neither shear nor double shear steel proving satisfactory to Huntsman, he tried melting shear steel in a clay crucible in a coke fire. This gave him a twofold advantage. Slag, which was present in the wrought iron and had remained in the shear steel, melted out and could be skimmed off the top of the molten steel in the crucible. In addition the carbon spread itself much more uniformly through the molten metal which, when cast into an ingot, was good carbon steel throughout. His success was not achieved without difficulty. In particular, he had to develop a crucible which would stand up to the heat required to melt steel.

Huntsman moved to Sheffield and set up a little crucible steelworks there, and this became the basis of Sheffield's present day world fame as a producer of high-grade special steels. Crucible steel is no longer made (though it did not die out finally until after the last war) but it was of great importance in the nineteenth century. The process was small-scale and the product consequently expensive, so crucible steel was used only where it was essential, as in engineers' cutting tools and for facing the cutting edges of such things as edge tools.

Wrought iron (wrongly called by some people malleable iron) and carbon steel, then, were the two metals which Bessemer set out to make by an improved method. In the end he made a material which superseded wrought iron, and was the forerunner of what are today called tonnage steels (the greater proportion of all steels made—principally mild steel but including the low carbon ranges). Bessemer was led to his famous invention by a series of experiments on a new type of rifled gun barrel which he had invented when the Crimean War broke out. It might be noted, in passing, that the steelmaking process was far from being Bessemer's only invention, for he was really a professional inventor and, by 1856, had to his credit successful inventions in such widely differering fields as glassmaking, textiles, sugar and non-ferrous metals. He was already, at the time of his most famous invention, quite a wealthy and successful man, but he was by his own admission neither a metallurgist nor even a practical ironmaker.

Some experiments on the melting of pig iron and blister steel together led him to try blowing air on to the surface of molten iron in a crucible and then he tried out the idea which was to lead to success—he blew air *through* molten iron. Bessemer's first little converter, tried out at his bronze-powder factory at St Pancras, London, held only 7 cwt of molten iron, but it was sufficient to show that the process worked. Indeed, it worked so well that Bessemer was rather alarmed at first. As the air was blown through the molten metal, by means of holes in the bottom of the vessel, flames issued from the vessel mouth, then large quantities of sparks came out and finally slag and molten metal were thrown out violently. All Bessemer could do was to watch and wonder what would happen next, for the valve which controlled the air blast was close to the converter, and nobody could approach it to turn it off. What did happen was that after a few more minutes the flame dropped. the eruption ceased and the molten metal lay quiet. When the metal was tapped and cast into an ingot, it was found to be wholly decarburized. Bessemer had made what was really a form of wrought iron, but without the mixture of slag which characterized the older product.

He patented the process before he announced it to the world (Bessemer was a first-class businessman as well as an inventor) and invited applications for licences to use it. There was naturally some scepticism. That cold air could be blown through molten iron without cooling it takes, on the face of it, a bit of believing. In fact the metal, at the end of the process, was hotter than at the beginning, for the oxygen in the air had reacted with the carbon in the iron to decarburize it, and this reaction is exothermic, that is it gives out heat. The oxygen reacted with the silicon in the iron, too, and this reaction is also exothermic.

A number of the larger ironmasters were ready to give the Bessemer process a trial, even if others remained suspicious, and several licences were taken out within a few days of the reading of the celebrated paper. Among the first works to set up Bessemer converters were the large Dowlais and Ebbw Vale Ironworks, in South Wales and Monmouthshire respectively. Foreign firms also took out licences. Apparatus was set up very quickly, the licensees made metal according to the inventor's instructions, and in every case the results were a failure. Bessemer had made perfectly good iron at Baxter House but all the licensees could do was to produce a brittle metal resembling the worst possible puddled iron.

Obviously there was something seriously wrong and Bessemer tried hard, and spent a lot of money, in vain, to discover the cause of the disaster. It was only after a lot of investigation that the truth was discovered, and even then for a time the process was only partly successful. When Bessemer had ordered pig iron for his Baxter House experiments he had, by the purest of chances, been sent some made by the Blaenavon Company in Monmouthshire. It was an iron which was remarkably free from phosphorus, and in this it was quite unlike most British pig irons. By chance again, Bessemer had lined his converter with non-siliceous refractory. Under these conditions no phosphorus passed into the finished metal. But at Ebbw Vale and Dowlais the available pig irons were phosphoric, and the converter lining materials were siliceous or, chemically, acidic. Phosphorus is not removed under acidic conditions, so it went into the finished metal. It will be remembered that phosphorus in iron causes it to be cold-short; this is what happened to the metal made by the licensees.

It was one thing to discover the cause of the trouble and another to remove it. Bessemer tried to find means of getting the unwanted element phosphorus out of the pig iron, but without success, and he was forced to the conclusion that the only thing to do was to use a pig iron which was phosphorus-free from the start. There were some such irons available in Britain—the hematite irons—and these, which were quite successful in the converter became known as 'Bessemer iron'. But this was only a part answer, for it left the major supplies of British and foreign irons, which were phosphoric, unusable in the Bessemer converter.

There was another trouble, too. Oxygen was absorbed very readily by the iron, and it had a bad effect on the finished metal, which would not settle in the ingot mould because of the large amount of gas evolved on cooling. R. F. Mushet (whose name has been mentioned before) found the answer to this problem, after being shown a sample of metal made at Ebbw Vale. Mushet removed the surplus oxygen by a deoxidizing agent known as spiegeleisen, a triple compound of iron, manganese and carbon. The manganese in the compound had a powerful affinity for oxygen, and the carbon was useful in controlling the final carbon content of the finished, or blown, Bessemer metal.

So, for a time, the difficulties of the Bessemer process were solved in part. Bessemer's reputation had naturally suffered and far from getting new licensees he found that some of the existing ones had lost interest. He considered that the only way to run the Bessemer process was to do it himself, and he set up a business in Sheffield as Henry Bessemer & Company in 1858. In time a few others came into the fold. John Brown started making Bessemer steel rails in Sheffield in 1860; Charles Cammell followed with the same product in 1861, also in Sheffield; Daniel Adamson made the first Bessemer steel boiler in 1860 and in 1863 entered into partnership to build a Bessemer steelworks at Penistone, Yorkshire. Bessemer steel rails were laid at Crewe station in 1863, and, as they proved successful, the LNWR set up its own Bessemer steelworks at Crewe in 1865. Some other works also took up the Bessemer process, in spite of its still being limited in scope by the necessity for using hematite irons, but official recognition of the metal was slow in coming. The War Office, it is true, accepted it in 1863, but the Admiralty would not allow its use until 1875, nearly twenty years after its original introduction.

Though the Bessemer process was in use by the 1860s it was still not really a very serious competitor to wrought iron. There were, however, other developments in the offing, one of which was a new steelmaking process, while the other was an improvement of the

Bessemer process itself, which was to strengthen its position enormously. Before considering these a short digression is necessary to mention some different developments which, while having no connection with bulk or tonnage steelmaking were nevertheless of importance and significant for the future. In 1864 Dr John Percy (1817–89) published his famous *Metallurgy of Iron* which was recognized as a model scientific treatment of a subject which had had all too little attention in this direction. It was the forerunner of many books of its type.

A year earlier, in 1863, Dr H. C. Sorby (1826–1908), an amateur scientist of considerable ability, had done his first important work, in his private laboratory in Sheffield, on the microscopical study of iron and steel. This aroused virtually no interest at the time, but it is now recognized as marking the beginning of the modern science of metallurgy. Scientific control of iron and steelmaking had taken a small but significant step forward.

The other development was that of self-hardening steel, by R. F. Mushet. He was still running a little crucible steelworks in the Forest of Dean, when he was asked by a Scottish manufacturer to make a new type of hard-metal tool. Finding this metal unsatisfactory he devised one himself. This was a completely new departure, having as an alloying element tungsten, and was the forerunner of a long line of alloy steels. Tungsten alloy steel had a special property; it would harden itself. Carbon steels could be hardened and tempered by suitable heat treatment, but if they were heated again they became soft once more. Tungsten alloy steel merely had to be forged to the required shape and allowed to cool. Then it was ground at the working point and it was ready for use. It could take very heavy cuts and did not soften like carbon steel if it got hot while cutting. This property was of great advantage to engineers who at the time were developing heavier machine tools which could make good use of Mushet's special steel. It came into use in 1868.

To return to the field of steelmaking generally, the development which followed close upon Bessemer's—the open-hearth process—is the next thing of importance to be considered. It was based on the work of C. W. Siemens (1823–83) a German who assumed British nationality and was knighted in 1882. Siemens was scientifically trained and applied his training systematically. His work, which led to the development of the open-hearth process, was at first directed to the improvement of furnaces—any furnaces, that is, not specifically steel furnaces. The first of the improved Siemens furnaces were used, in fact, in the glass industry.

Siemens was concerned primarily with fuel economy, and to this end he applied the regenerative principle of waste heat recovery. We have already met two methods of recovering waste heat; Rastrick's waste-heat boiler, and the boilers and stoves in which blast furnace gases were burnt. Both these were really of the recuperative type, in which heat is exchanged continuously between hot gases and water or air. Siemens took his hot gases through a honeycomb mass of firebrick, and, when the brick was very hot, cut off the flow of gases and let air pass through the hot mass to collect its heat. By having two masses of firebrick, or regenerators, Siemens heated one while the other gave up its heat, and reversed the hot gas and air flows at intervals to keep up a continuous flow of hot air, which he passed into the furnace for fuel combustion. Very high air temperatures could be obtained by this method, and Siemens was already recovering so much heat in 1857 that he claimed to save 70 to 80 percent of the fuel used on an orthodox furnace.

At first the principle was applied to coal-fired furnaces, but ash from the coal was carried into the regenerators, where it caused choking, and Siemens soon introduced a gas producer in which he gasified the coal before burning it, as gas, in the furnace. This solved the problem of clogging the regenerators and at the same time made it possible to use a cheaper fuel, small coal. Siemens was successful with his regenerative furnace in some industries, notably glass-making but not, at first, in steel.

While Siemens was trying to interest British steelmakers in his new furnace, the regenerative principle was applied in 1857 by his friend E. A. Cowper, to the hot-blast stove for blast furnaces. Cowper used an iron shell lined with firebrick and filled with honeycomb bricks, burnt blast furnace gas in it until the honeycomb was very hot, then cut off the gas and sent the blast air through it. This enabled the blast to be heated to very high temperatures indeed, and gave further economy in blast furnace operation. In practice three or more stoves were used to each furnace, with two or more heating while one was giving up its heat to the blast. The Cowper stove was highly successful, and is basically the standard hot-blast stove of today throughout the world.

In 1863 the Frenchmen Emile and Pierre Martin, who had taken out a licence from Siemens, used one of his furnaces for making steel; from this arose the practice, formerly common, of referring to the open-hearth process as the Siemens-Martin process. This should be discouraged, for the Siemens-Martin process is really only one of several which can be carried out in the open-hearth furnace; its peculiarity lies in the fact that a proportion of the metal charged is scrap. In 1866, finding no support from British firms, Siemens rented a small factory in Birmingham, called it the 'Sample Steelworks', and experimented himself with steelmaking in his new furnace. A year later he took out a patent for making steel in the open-hearth furnace, and in 1869 the Landore Siemens Steel Company started its works at Swansea, where, at first, it made 75 tons of steel a week in the open-hearth. Progress was slow at the start, but other works followed, and by the 1870s open-hearth steel was well established. The process became eventually of major importance in Britain and some overseas countries; and is only now losing its pre-eminence under competition from completely new processes.

The process itself is fundamentally simple, though it is broadly adaptable, and its chemistry varies according to how it is used. An open-hearth furnace is not unlike a puddling furnace inside, though bigger. Today it is very much bigger; furnaces exist which can hold 500 tons of iron. It has a bowl or hearth in which the pig iron and/or scrap to be melted is placed (nowadays mechanically) and is fired by coal gas (today often by oil fuel) alternately from opposite ends. Waste gases pass out of ports in the furnace end walls and enter chambers containing chequer firebricks, to which they give up a proportion of their heat. Finally, the gases go to a chimney stack. Sometimes, today, they pass through a waste-heat boiler on the way to the stack but in the early days of the furnace such boilers were not used.

The regenerator chambers are in duplicate, one set at each end of the furnace, and by reversing the direction of firing they are alternately used to give up heat to the combustion air and to receive heat from the waste gases. In practice there were on the Siemens furnace two extra sets of regenerators which were smaller, and were used for heating the gas for firing the furnace.

Iron to be made into steel is charged into the furnace and there brought up to the required temperature. It can be charged either as cold pig iron and cold scrap, or as cold scrap alone, or the pig iron can be charged molten from a ladle. At the same time the slag-forming or fluxing constituents are charged; these will form a slag which does the decarburization of the iron. They will vary according to the analysis of the original metal and to the details of the process, and may include limestone, fluorspar (calcium fluoride, CaF_2), millscale and iron ore.

When the furnace has been charged with the required raw materials (iron and scrap) and fettling (fluxing materials) the melting period commences, and gradually, during this period, some of the unwanted elements in the charge pass into the slag. When these elements have been cleared, the period known as refining begins, and the molten contents of the furnace are subjected to a process of controlled oxidation to bring the carbon down to the required figure.

Open-hearth steelmaking did not replace the Bessemer process; it supplemented it, for the two processes, though having the same end product, worked in different ways and could be varied according to the raw materials. Of the two the open-hearth process was the more adaptable to circumstances. It was slower in operation, and this could be used to advantage. To work a heat in the open-hearth furnace took six, twelve or even fifteen hours, and it was possible to take a sample of the steel, determine its carbon content, and stop the process by tapping the furnace just when the carbon was at the required figure. On the other hand the Bessemer converter works fast, making a charge of steel in about half an hour, and the normal procedure was to blow the charge until all the carbon was out, and then recarburize to the required figure by adding a carbon-bearing material in the ladle into which the metal was tapped. Spiegeleisen, previously mentioned, contains carbon, and by adding the appropriate quantity the necessary carbon content could usually be obtained. If more carbon were needed, it could be provided by adding anthracite coal dust to the ladle.

From the earliest days the Bessemer converter was mounted on trunnions so that it could be tilted for the molten iron to be charged, and again for the blown metal to be tapped off. The first open-hearth furnaces, on the other hand, were fixed, and tapping was by driving out a refractory plug which stopped up a hole at the bottom. In later years some open-hearth furnaces have been made so that they can be tilted sideways to assist in tapping, but this was not done for many years after the process became widespread.

By the 1870s steel, made either by the Bessemer or the open-hearth process, was well established, but the Bessemer process still suffered from the need to use low-phosphorus iron. In the 1880s the position changed, for at the beginning of that decade a modification to the process enabled phosphoric irons to be used without difficulty and without restriction. This modification was the work of Sidney Gilchrist Thomas (1850–85), whose knowledge of the chemistry of steelmaking, acquired by intensive study in his spare time, enabled him to discover how to eliminate the unwanted element phosphorus in the converter itself. Thomas, whose family circumstances compelled him to earn his living as a police-court clerk in London, was studying chemistry when he heard it said that the man who could eliminate phosphorus in the Bessemer converter would make a fortune. This proved to be true; Thomas did make a fortune and his modified Bessemer process was adopted all over the world, particularly in some European countries where it suited the local pig irons. His experiments were conducted under great difficulties, his time was limited, and his apparatus scanty, and he solved the problem in theory before he got an opportunity to carry out full-scale trials at an ironworks.

A knowledge of chemistry told Thomas that when phosphoric iron was blown in the Bessemer converter the phosphorus oxidized rapidly to form phosphoric acid, and if the lining of the converter was also acid (in the chemical sense) the phosphorus would stay in the iron. Two acids have no affinity for each other. But a basic material (again in the chemical sense) would have an affinity for phosphoric acid, would combine with it, and would take it out of the iron. So he tried various materials for basic linings, and at last, with the aid of his cousin, P. C. Gilchrist, chemist at Blaenavon Ironworks, Monmouthshire, was able to make a proper trial of his ideas.

Naturally there were problems at first, but in 1879 Thomas resigned his police-court job and devoted his full time to his converter work. Success then came quickly. In his successful experiments Thomas used dolomite, a form of limestone, as his converter lining material, and this was the basis of what became widely known as the Thomas process (particularly in Europe) but is better described as the

17 A two-high merchant rolling mill of the late 18th century. From Wortley Low Forge, near Sheffield, now preserved at Wortley Ironworks by the Sheffield Trades Historical Society

18 Hand rolling of sheet

19 (*above*) Part of an original Bedson continuous rolling mill, showing the alternate arrangement of rolls

20 (*left*) The last cementation furnace to work in Sheffield (closed 1952) at the works of Daniel Doncaster & Sons Ltd

basic Bessemer process, in distinction to the original Bessemer process, which is acid. The first licensed user of the basic process was the firm of Bolckow, Vaughan & Company, Middlesbrough, and the first steel was made there in a basic converter on 4 April 1879. Basic steel spread widely, though not to the exclusion of the acid process, which continued to find favour where the available pig irons suited it. Basic, however, became by far the biggest producer. It is only in recent years that the two types of Bessemer process have given way to newer methods of manufacture, and even now they are not extinct.

The basic process of steelmaking was not confined to the Bessemer converter. Basic chemistry was equally applicable to the open-hearth process, to deal with high-phosphorus irons, and basic steelmaking became, in fact, a major feature of the British iron and steel industry. Since the phosphorus passed into the slag Thomas ground the cold slag into a powder and sold it as an agricultural fertilizer; this practice, too, developed and has continued.

Thus, by the 1870s British (and, of course, foreign) steelmakers had at their disposal two proved processes, the Bessemer and the open-hearth, either of which could be adapted to a variety of pig irons and other raw materials, and both of which were capable of large-scale operation. For special and alloy steelmakers there were as yet no revolutionary processes, but for the industry as a whole there were, besides means of making tonnage steel, new ways of processing it by rolling.

These had developed coevally with the new steelmaking processes, to which they were complementary. Bigger-scale steel production needs bigger-scale rolling processes and vice versa; neither is much use without the other.

In 1861 Sir John Alleyne (1820–1912), of Butterley Ironworks, Derbyshire, had made a start with the mechanization of rolling mills, thus enabling heavier pieces to be rolled. He patented mechanical traversers for moving the rolled piece both in front of and behind the rolling mill, and thus made it possible for one man to manipulate quickly and easily pieces of hot iron which would otherwise need the manual effort of many. Alleyne was concerned with heavy structural sections (he was associated *inter alia*, with H. W. Barlow in the construction of St Pancras station roof), and devised more than one means of making them; his traverser was the most important. It paved the way to the production of much larger joists and channels and came at just the right time, when larger pieces of metal (mild steel) were available. Alleyne also designed and patented a new type of rolling mill, in which the direction of rotation of the rolls was reversible at will. Thus, a piece could be rolled in both directions, as in a three-high mill, but without the need for lifting it too high on the return pass. Such a mill was the ideal complement to the mechanical traverser.

The Alleyne reversing mill, however, though its conception was right, was not a particularly good design mechanically, and it was left to John Ramsbottom (1814–97), of Crewe railway works, to introduce the first true reversing mill. In 1866 he took the engine part of a railway locomotive and coupled it to a two-high rolling mill. By means of the ordinary link-motion reversing gear the mill driver could run the rolls in either direction as required, and, since the reversing gear was worked by a simple lever, the mill could be reversed quickly. Ramsbottom's reversing mill and Alleyne's traverser were the basis of the heavy mechanized mill which was to become of great importance in the latter part of the nineteenth century, though of course the detailed design of both underwent many changes over the years.

In the rolling of lighter products, too, there were important developments, both of which enabled the process to be speeded up. The first of these was the looping, or Belgian mill, which was based on an idea of some unnamed Belgian in about 1860. In an ordinary three-high mill the piece was allowed to run right out of the rolls at each pass, then caught manually with tongs and led to the next pass. This was quite satisfactory as long as the piece was short (it remains the practice today in such circumstances) but where the piece was long and slender, as in the smaller sizes, a lot of heat was wasted. The Belgian idea was to catch the end of the piece with tongs almost as soon as it left the rolls, run round with it through 180 degrees and enter it into the next pass immediately. The piece was thus, for a time, in two passes simultaneously, and this saved time and heat. Looping was applied quite widely to the rolling of small iron rods and bars in Britain, and was also applied, later, to steel just as successfully. Looping or Belgian mills are in use today, though their number is small, much of their work having passed to more highly mechanized equipment. In some of the modern mechanized mills looping is practised, but it is done mechanically by means of what are called repeaters, instead of manually.

The other development in rolling mills for light products in the 1860s was much more revolutionary. It was the continuous mill. Charles While, of Pontypridd, patented such a mill in 1861. This, however, was not brought into commercial operation, and George Bedson, of Manchester, was the man who really made the continuous mill work. Bedson was interested in small-diameter iron rod for drawing into wire, and he introduced two novel features in his design, which was put into operation in 1862.

In the Bedson mill there were several stands arranged in tandem, and only one pass was made in each. The first of the stands was placed as close as possible to the furnace in which the iron billets for rolling were heated, and a billet was pushed out into the first stand. On emerging from this pass the piece went straight into the next stand, which was fixed as close as possible to the first one. From the second stand the piece went straight into the third, and so on down the line of stands until it was rolled to the required size. Bedson's first mill had sixteen stands in tandem. This was the first entirely new feature.

The second was in the arrangement of alternate stands with the rolls horizontal and vertical. By this arrangement work was done on the piece alternately at 90 deg, and a sequence of rollings from oval to round could be carried out as required without any need to turn the piece itself between passes. There was therefore no risk of scratching or other damage to the piece which was always possible when it was turned through 90 deg, and the product was free of fins, an essential requirement of the wire drawer. As the piece was being rolled in several stands simultaneously (it could be entering the last one before it had left the first), and the stands were close together, neither heat nor time were lost and the mill was ideal for rolling any small bars, rods or sections.

It was not easy, at the time, to arrange the mill drive. A piece of metal being rolled always emerges from the rolls faster than it goes in; this is inevitable, since it is getting longer. In a mill where the piece emerges fully from one pass before it enters the next, this increasing speed does not matter. In a looping mill, the loop itself acts as a 'reservoir' of metal between the passes. But in a tandem mill like Bedson's, each successive stand has to run at a higher speed than the preceding one. In Bedson's time the only practicable source of power was the steam engine, and the only way to drive the 16 tandem stands was to have an engine-driven shaft running alongside the mill and take the drive to each stand through gears. In practice Bedson used bevel gears, each successive bevel on the shaft being bigger than the preceding one, and so driving the stand faster. This involved calculating the speed increases from stand to stand and designing the gears accordingly, which was at the time no mean achievement.

Bedson's mill worked admirably in the Manchester works, and a second one was built a few years later; this one ran until the present century. Elsewhere the design was only adopted slowly, as it had a fairly limited application and was, by comparison with a hand mill, costly. In the 1860s the main market for small rods was the wire trade; not many other users could take large quantities of one or two sizes of rod, which the Bedson mill was most suited to produce. In the jobbing trade, where small quantities of many sizes and shapes were needed, the conventional hand mill could hold its own as, indeed, in a limited way, it continues to do even now. The real significance of the Bedson mill was that it established and proved a principle which was to have widespread application in later years and is now applied to other products besides small bars and rods, as will be shown later.

But the Bedson mill was not only used in Britain. A mill was exported to America where the man in charge made some improvements and eventually set up as a manufacturer of continuous mills to the trade. From this circumstance has arisen the common belief that the continuous mill was an American invention. It was not. George Bedson invented the principle and put it to work successfully and it is only fair to his memory to make this fact clear. Of course, changes in detail were made from time to time by Bedson's successors, one of them being the abandonment of the vertical/horizontal roll arrangement in favour of mechanical guides which twisted the piece through 90 degrees between passes. It is of interest to note that in the latest designs of continuous bar and rod mill, the original Bedson arrangement has been revived.

So by the late 1860s the iron and steel rolling side of the industry was well on the way to mechanization. Further developments and new products were to follow fairly rapidly. Before these can be considered, however, it is necessary to take a look at a new method of working the blast furnace, which came in at about the same time as the rolling processes described above. This was working with the closed forepart, which has received scant attention from historians and yet was of considerable importance, for with the arrival of the closed forepart the blast furnace had almost reached the form used everywhere in the world today. It was in fact, with the exception of the mechanical charging gear which was developed later, identical in form with present furnaces; the present ones are very much larger, that is all.

The closed forepart was not a British invention, though it was adopted on a small number of furnaces soon after it was patented in Britain, on behalf of the inventor, F. Lührmann, of Osnabruck, Prussia, in 1867. It will be recalled that the front or forepart of the blast furnace was built with a space under the tymp which was closed by clay and an iron plate when the furnace was blowing, and opened during tapping. Lührmann simply made the furnace hearth the same all the way round, dispensing with the tymp altogether. He arranged a tap hole and slag notch in the solid wall of the hearth, and closed them in the normal way. This simple rearrangement of the forepart made running the furnace easier and tapping quicker, and paved the way to the higher blast pressures which were to come in due course. The blast pressures used today would be impossible with an open forepart. Fletcher Solly and Urwick installed a closed forepart at Willenhall Furnaces, Staffordshire, in 1869, and others followed slowly. The open forepart lingered on until recent years at a few small plants, but it has long since ceased to be used on any modern furnace.

CHAPTER SEVEN

New Products and the Impact of Electric Power, 1890 - 1945

Developments came rapidly at the end of the nineteenth century. Some were internal to the industry, such as new plant and techniques; others were the result of external influences, such as the introduction of electric power. There were important changes introduced from abroad, too, for some overseas countries had been very busy extending their own iron and steel industries, and several of their ideas were well worth adopting, even though conditions in Britain were often very different from those where the imported ideas originated.

It was towards the end of the nineteenth century that technical initiative passed, for the first time, to America, and some of the labour-saving devices adopted in Britain came from the U.S.A. This does not mean that the initiative passed wholly to the U.S.A. or that it stayed there exclusively. On the contrary, Britain still had some valuable contributions to make, and so had other countries. Steelmaking was becoming, from a technical point of view, international, a trend which has developed until at the present time a modern steelworks anywhere is based on a host of machines, processes and ideas from many parts of the world.

But in the latter part of the nineteenth century America had a peculiar problem, shortage of skilled manpower in a developing economy, and it solved the problem in the only logical way. It put machines to do the work of men. This it did, of course, in many other fields besides that of iron and steel manufacture, but we need not concern ourselves with other industries except to say that the more America mechanized, the greater was the demand for iron and steel and the greater, therefore, the pressure to mechanize there.

By 1890 the U.S.A. had become the world's largest producer of iron and steel (which it still is), American works were much bigger, and individual units within the works were getting too big for men to handle without extensive mechanization. Moreover, American plant was made to work harder; the idea of fast 'driving' of the blast furnace, which is now very widespread, first took root there. Thus outputs became greater and the quantities of raw and finished materials in a specific space of time grew accordingly.

In 1895 a machine was invented in the U.S.A. which, in time, was to render the older sand pig-bed at the blast furnace obsolete. This was the pig-casting machine. Molten iron from the furnace tap hole ran down a spout into a moving, endless chain of shallow moulds, where it cooled quickly, aided by water sprays, into relatively thin, cushion-shaped pigs. After travelling a short distance the moulds turned upside down automatically, the pigs, by then solid, fell out into a waiting wagon, and the moulds returned to the head of the machine for re-use. The advantages of such a machine are very obvious if it is related to a furnace making, say 2,000 tons of iron a week, which several in Britain were capable of doing in the 1890s. This represents about 300 tons/day (as production was continuous), and as the furnace was tapped two or three times a day, anything from 100 to 150 tons of pig iron would have to be cleared manually from the pig bed. As the pigs weighed about 1 cwt each, something like 2,000 or 3,000 of them would need lifting and carrying, albeit only a short distance, at every cast. Of course, not every British furnace was as big as this, at that time, but some American ones were bigger, and the handling problem would have been virtually insoluble without the pig-casting machine. Some such machines had been installed in Britain before the end of the century.

Fast driving of a blast furnace brought into use another American idea, the tap-hole gun, or clay gun. This was not unlike a cannon in appearance, and using steam (at first, later compressed air) as the motive power, shot a plug of clay into the tap hole to close it after tapping. With manual tap-hole closing it was necessary to take the blast off the furnace for a man to approach the forepart; this wasted time. With the tap-hole gun no time was lost, the tap hole was closed in a matter of seconds without taking the blast off at all. Again, some tap-hole guns soon found their way to Britain.

But the most important American invention to be imported into Britain in the period under review was the mechanically charged blast furnace. Without this no really large furnace could operate.

Consider once again our hypothetical furnace making 2,000 tons of iron a week. It would need at least 4,000 tons of ore and quite likely a lot more, in the same period, plus coke and limestone; at the very least some 5,000 tons of materials would have to be handled manually out of railway wagons or canal boats into charging barrows, wheeled to the furnace incline or lift, and then tipped into the furnace throat. In a land where furnaces were getting bigger and labour was at a premium, it is hardly surprising that means were devised to do the charging mechanically.

Charging a furnace mechanically involved a change in the bell and hopper which, it will be remembered, dated from 1850. Mechanical charging was done by a self-emptying skip which was hauled up an incline to the furnace top and discharged there into a receiving hopper. Below this hopper was a revolving distributor and below this again was a bell—the small bell. Finally, below the small bell was a second one, the lower or big bell. All this was necessary because the charges have to be carefully distributed around the interior of the furnace. Failure to do this could result in uneven working of the furnace. When it was charged manually the men wheeled their barrows to different points at the furnace throat and tipped them to set up a pattern of distribution. A mechanically charged furnace, however, had no men at the top, so the distributor, which was rotated in angular steps mechanically by remote control, was introduced. Charges of ore and coke were loaded into the skip from bunkers, hoisted and deposited in the distributor, which placed them on the small bell in the required pattern. When a sufficient charge had been deposited on the small bell, it was lowered, and the charge fell on to the large bell. This held a considerable quantity of material, and was lowered at less frequent intervals to allow the charge to enter the furnace. The furnace top was now completely closed, and no gas was lost to the atmosphere when the big bell was lowered for charging; the small bell retained it.

Although the mechanically charged blast furnace was an American invention, Britain was not far behind in adopting the idea. The first such furnace in this country was built at Frodingham Ironworks, Scunthorpe, Lincolnshire, in 1904–5 and this was also probably the first of its kind in Europe.

Thus, by the end of the nineteenth century, blast furnace operators were able to use very much bigger furnaces, with closed foreparts, mechanical charging, mechanical closing of the tap hole after tapping, and mechanical pig-casting. Blast temperatures were much higher, too, and the rate of driving was such that the outputs went up out of proportion to the increased size. Not all the furnaces in Britain went over to these new methods, of course; some completely new ones were built for hand charging and quite a number of the existing ones continued to be hand charged and to use the old-style pig bed. But the pattern had been set for the large furnaces of today, some of which have hearths over 30 ft in diameter, are well over 100 ft high, and produce as much as 2,500 tons of iron a day. Some work had also been done in America, in 1906, on preparation of the charge materials for the blast furnaces, when A. S. Dwight and R. L. Lloyd devised a means of using the dust and fine ores which would otherwise have been unusable, since they would have choked the furnace. Dwight and Lloyd subjected fine ores to a process of controlled burning with small coke or breeze, so agglomerating or sintering them into a semi-fused cake which could be broken easily into lumps of a suitable size for the blast furnace. This process was to become, in time, of considerable importance, and much useful development work on it was done in Britain. This, however, will be referred to later in the correct chronological sequence.

It was not only at the blast furnaces that the end of the nineteenth century and the first part of the twentieth saw great changes. Bulk or tonnage steelmaking underwent little change for some time following the successful development of the Bessemer and the open-hearth processes, but there were developments in steel rolling, particularly of the heavier products. In addition, electricity began to be used in iron and steelworks, and this was a portent of many great changes to come. Electricity arrived in the mills in the 1880s, at first on a very small scale, and mainly for light power applications and for lighting. But having found its way into the works electricity soon showed its potential as a controllable source of power. It was very convenient, for an electric motor could be put anywhere, needed little space and only simple foundations, and could be connected to the source of power by a simple run of wiring. Its potential was actually very much greater than at first sight appeared, for in due course it showed itself to be adaptable to very sophisticated control devices, and these are the basis of many complicated mills of today.

In steel rolling the first major change was the introduction of the universal mill, in which four rolls acted simultaneously on a piece being rolled, instead of the traditional two. It was first applied to steel plate rolling, with the object of obtaining good edges on a plate and better dimensional accuracy in width. Although it was a British invention, it was first used in Europe. Nevertheless, it was applied in Britain well before the period at present under consideration, the first universal plate mill in this country being installed at Britannia Steelworks, Middlesbrough, in 1878. This pioneer installation, however, was not without its troubles, and it was another ten years before a second universal plate mill was put to work. This was at Blochairn Steelworks, Scotland, and the design was modified in that all four rolls were individually power driven. Again, there were mechanical troubles and the mill was not a great success operationally, but, with the earlier example, it had pointed the way to new developments.

The next universal mill was very successful, and it was with this mill, which started work in 1902, that universal rolling really started on a commercial scale. It was a British invention, the work of Henry Grey (1849–1913) in 1897, but nobody could be found to try it out until 1902, when a Grey mill was installed at Differdingen in Germany. Grey's design was quite different from the earlier mills and so was its product, for it rolled beams, not plates. This was more difficult to do and called for some good engineering design work.

Beams are shaped roughly like a letter H, (they were often called 'H beams') and to roll them on all sides at the same time meant that there had to be two rolls of more or less conventional pattern and two more, at right angles to them, to roll the outsides of the beam flanges. These rolls had to be short and stubby. For mechanical reasons it was impossible for all the rolls to bear on the beam in exactly the same plane, but they did so nearly enough to get the desired effect. In practice, the beams were first 'roughed' out in normal stands from blooms into the shape of a letter H, and then given two finishing passes in two separate, true universal stands. This is still the practice in modern universal mills, but of course the controls and accuracy are better, and the drive is all-electric. Grey's mill was steam driven, but it did the job it was intended to do, and worked for many years. Only recently, in fact, has it been replaced by a modern mill of British manufacture. Curiously, though the universal beam mill was quite successful, and the product it made superior to the ordinary joist, the idea took a long time to get established, and it is only in recent years that universal beams have become, in the other sense, 'universal'. Today all the beams rolled in Britain are universal beams. A British Standard (BS 4:1962) covers these beams, and the older joists, which are of a slightly different shape, having flanges tapered on the inside, have been withdrawn. It is not without interest that the withdrawn range of beams was the first standard issued by the then newly formed (1901) Engineering Standards Committee in 1903; this committee became the British Standards Institution.

Products, as well as processes and plant, developed in the last few years of the nineteenth century and in the first part of the twentieth. Alloy steels, in particular, came into prominence. Alloy steelmaking really goes back to Huntsman who, as has been said, devised in 1740 a means of making a consistent hardenable steel, but this, carbon steel, is not really an alloy in the modern sense of the word. Mushet, whose work has also been referred to, became the 'father' of alloy steelmaking when he introduced his 'Special Steel' in 1868. After that there was nothing of importance to record until Hadfield developed manganese steel in 1887.

Robert Hadfield (later Sir Robert, 1858–1940), was one of the outstanding metallurgists of his time. He was the son of a Sheffield steel founder, and his detailed study of steel arose out of his attempts to improve the quality of the family firm's products. Prior to Hadfield's invention of manganese steel the principal characteristic of alloy steel (and the only one really sought after), was the fact that it could be hardened. Manganese, when alloyed with a carbon steel, gave it some interesting new properties. Hadfield's first manganese steel contained 12·5 per cent manganese and 1·2 per cent carbon. It was not hard but was very tough and capable of resisting abrasive wear. This property could be improved by suitable heat-treatment. Hadfield also discovered that when a manganese steel contained more than 10 per cent manganese it was non-magnetic. This was not of great importance at first, but came to be so in due course, as electrical science developed.

Toughness and wear resistance were the most useful properties of manganese steel at first, and it was used for railway points and crossings, giving a life of five or six times that of ordinary steel. It was (and still is) used, too, in such machines as rock crushers. Manganese steels are still used widely for applications demanding toughness, and although some improvements have been made in heat-treatment procedures, the alloy itself remains largely as Hadfield first made it. Manganese steel was not Hadfield's only discovery. He was also associated with the development of silicon steel, which is highly magnetic and in this respect the reverse of Hadfield's other alloy. Silicon steel again was of comparatively little importance at first, but is now widely used by electrical engineers.

It was in the early years of the twentieth century that the best known development in alloy steels took place. This was the introduction of stainless steel, and it was the work of a Sheffield man, Harry Brearley (1871–1948). Brearley had been at work for some time on improving steel for rifle barrels when, in 1913, he made a steel containing 12·68 per cent of chromium and 0·24 per cent carbon. His object was to produce a steel which would be free from fouling and erosion but he found that the steel had a novel property. To make a microscopical examination of it he had to polish and etch the surface, and the new steel would not etch with any of the usual acids. Brearley knew immediately that if the steel would not etch it would not be affected by at least some, if not all, the causes of corrosion, since etching is really only a form of corrosion.

He had, in fact, made stainless steel, or at least the first of the great family of stainless steels which are so widely used today.

Living and working in the great cutlery centre, Sheffield, Brearley thought of persuading a local cutler to make some knives of his new steel, and this he did after some difficulty, for people found it very hard to believe that any steel could be corrosion-free. Nevertheless, it was. Knives made from it were unaffected by water and by all the mild acids met with in food and cookery.

There were problems at first, for stainless steel proved difficult to work, and the knives made from it were thick and blunt. Nor could the early stainless steels be hardened by heat treatment. The war of 1914–18, coming so soon after Brearley's development, prevented any further advances in the alloy, but it was taken up again later, and today it is very well known in the domestic field as well as in industry.

A whole range of stainless steels has grown out of Brearley's first one and today there are stainless alloys which can be hardened and tempered as well as others which are not affected by heat treatment. There are also stainless steels which are heat resisting.

In the period 1914–18 there was naturally little technical development, for the only thing that counted was production for war needs. But the war did bring about one important development, the extension of the use of the electric furnace. This was actually much older than the period under review, though it found little or no application at first. Sir William Siemens had shown that if an electric arc is struck between two electrodes, great heat is generated, and in 1878 he patented an arc furnace. He was followed in 1886 by the Frenchman, Paul Heroult, who used the arc furnace for the production of aluminium and later, in 1889, for making ferro-alloys. He took the development further when, in 1900, he made steel in the arc furnace.

In the 1914–18 war a problem arose in dealing with the turnings and other swarf from the machines making some of the munitions of war. This swarf was a valuable material if it could be remelted, but neither the open-hearth furnace nor the Bessemer converter could deal with it satisfactorily. The only conventional melting equipment which would melt swarf was the crucible furnace, and this was slow and of small output. The electric arc furnace provided the answer. It generated the necessary heat readily and since it used no solid, gaseous or liquid fuel in the furnace itself, there was no possibility of contamination of the steel by impurities in the fuel. Moreover, it was not just a means of melting scrap; it could be used for steelmaking as well.

Electric-arc steel melting and steelmaking developed quite rapidly during the 1914–18 war, particularly in the Sheffield area, which was then, as it is now, the principal centre of the alloy steel trade.

The other type of electric furnace, the induction furnace, which is now used extensively in steelworks, also dates from the nineteenth century—from 1877 in fact, when it was first made in Italy—but it took longer to develop, and the first one in Britain was installed in 1927. The induction furnace uses the same principle as an electrical transformer, in which an electric current is used to induce a current in a secondary circuit. In this case the secondary circuit is the metal to be melted, which is held in a refractory vessel surrounded by the primary winding. It is structurally simple, the whole furnace being mounted on trunnions so that it can be tilted forward for tapping, and is effectively a self-contained crucible with its own source of heat—electricity, though of course it is capable of much more accurate control than the old coke-fired furnace. It should be pointed out that the induction furnace is essentially a device for melting. Steelmaking ingredients are carefully weighed out to give the required composition and then melted in the furnace; no refining is done there, and the furnace does not make steel out of iron. But alloying materials can be added to the melt and the final composition can be altered.

The most important single development in steel processing to emerge before the 1939–45 war was again American. This was continuous wide strip rolling, and its American origin was quite logical. In the America of the 1920s there was a vast and expanding market for all types of mechanical devices, including one which was relatively new—the automobile. This brought demands on the steel industry for sheets for bodywork and other components which could not be met easily, if at all, by the conventional plant of the time. In addition there were other goods which demanded steel sheet, notably the so-called consumer durables such as cookers and, later, refrigerators. Then, too, there was a growing market for canned goods and this called for tinplate which, it will be remembered, is actually thin sheet steel with a microscopically thin coating of tin.

Hand sheet mills could not cope with the demands of the early 1920s, especially in a country where labour was scarce and dear. The American answer to this problem was to mechanize sheet rolling, and the plant developed was the continuous wide strip mill. This term requires some explanation, for it is not immediately apparent what strip (even 'wide' strip) has to do with sheet. But the name is quite descriptive, for what the mill produces is a very long strip of steel up to several feet wide, which is rolled continuously and at high speed, and subsequently cut up automatically into sheets of the required lengths. Today the wide strip is not always cut up but is often sent away in the form of coils weighing several tons and used in the customers' works on machinery specially designed to deal with steel in this form.

The world's first wide strip mill came into production in America in 1923, and the process has now spread to all major steel producing countries. Only a very small amount of steel sheet is now made in

NEW PRODUCTS AND ELECTRIFICATION, 1890–1945

hand mills, and most of this is special steel or is rolled for some special purpose.

In the continuous mill the steel, prepared in the form of a slab, is hot rolled in a series of tandem stands, one after the other in a straight line, to light plate thickness, and coiled mechanically at the end of the line. It is then cleaned in acid (that is, pickled), to remove the iron oxide or mill scale, and again rolled, this time cold, in a separate series of tandem stands, to the required sheet thickness. Finally, the coils are taken to a separate machine, the flying shear, and there cut, fully automatically, into sheet lengths. Alternatively, as already stated, the coils may be despatched as they are for use in customers' works. Other equipment used in continuous wide strip mills includes annealing furnaces, which take a stack of coils for heat treatment as required. Sometimes the steel is required 'dead' soft, as annealed, sometimes it will be given a single pass in a cold mill (temper-rolled or skin-passed) to give it the degree of cold working, or stiffness needed for the job it is to do. But the main part of continuous wide strip rolling is done in the hot and cold tandem mills.

For a time continuous wide strip rolling did not spread outside the land of its origin. It was a very successful process, but it was costly to install, maintain and operate, and could only be justified where there were very large markets for steel sheet. Such markets did not exist outside America in the 1920s. But the process had so many advantages that it was bound to spread in the end, and the spread began when, in 1938, the first continuous wide strip mill outside America was put to work at Ebbw Vale Steelworks, Monmouthshire, a works which had pioneered more than one historic development in iron and steelmaking.

The Ebbw Vale wide strip mill was designed to produce strip primarily for tinplate making, and its output of about 5,000 tons a week was far in advance of anything in Britain at the time, though small by the standards of today. Although strip for the adjoining tinplate works was the primary product of the mill it was quite capable of producing plain untinned or 'black' sheet, which could be sold as it was or galvanized and corrugated to form the familiar galvanized 'iron' (which it is still called often enough today, though it has been steel, not iron, for many years).

Ebbw Vale mill came in time to be of considerable value to Britain in the war years which soon followed its introduction, but there was no further development in British wide strip rolling while the war lasted. The war, however, called for some prodigious efforts on the part of the iron and steel industry, with the inevitable result that much of its plant was worked to death, and the need for re-equipping was pressing. But at first little could be done but plan developments, for the post-war boom in demand for steel of all kinds kept every producer in full swing. One thing which emerged from a close examination of the industry immediately after the war was that the producer of the future was going to be the big works. Plant was becoming so large, complicated and costly that firms were forced to amalgamate in order to have the necessary financial resources, and the small concern, except for certain specialized producers, was bound to become a thing of the past. Furthermore, many new developments were in hand or in the offing, and the process of integration was bound to accelerate. This is what in fact has happened, when we come to consider the industry as it is today.

Many temporary expedients had been necessary during the war to overcome, or at least to lessen the effect of shortages of some raw materials, but most of these arrangements were discarded when the materials once more became available. Some of them, after all, had been no more than poor substitutes and the best thing to do was to get rid of them as soon as possible.

One change forced on a branch of the industry by the war, however, was an improved method of making tinplate. This, developed in America because of a critical shortage of tin, was the electrolytic tinning process. It has stayed and spread because it makes a better product faster and cheaper. To understand electrolytic tinning it is first necessary to take a brief look at the old process it superseded. From the earliest days tinplate had been made by cleaning the sheets by pickling them in acid and then dipping them in a bath of molten tin, a layer of which adhered to the surface and cooled when the sheet was withdrawn from the tin 'pot'.

At first, tinning was done by hand. Then, in the nineteenth century a comparatively simple machine took over the actual work of tinning, the sheets being passed mechanically through the molten tin one after another. Some further mechanization developed in the ancillary processes, such as pickling and cleaning of the sheets, but by the outbreak of war in 1939 the whole process of tinning was technically far behind that of actually producing the sheets.

21 Four modern blast furnaces, the 'Iron Queens': 'Queen Mary', 'Queen Bess', 'Queen Anne' and 'Queen Victoria' at the Scunthorpe works of the Appleby-Frodingham Steel Company

22 A 110-ton electric arc furnace at the works of Steel, Peech and Tozer, Rotherham

23 A 25-ton Bessemer converter, blowing, at the Workington Iron and Steel Company, Cumberland

NEW PRODUCTS AND ELECTRIFICATION, 1890-1945

Wartime shortages of tin were felt in America, where the use of tinplate was highly developed, just as they were in Britain, and it was in America that a solution to the problem was found. This, the electrolytic tinning process, is akin to electroplating, but is done continuously as the thin sheet steel is unwound from a coil, passed through the tinning line and re-coiled at the other end. By this means a very thin layer of tin can be applied uniformly over the entire length of the strip and the first object, economy in the use of tin, is achieved. Electrolytic tinning, however, has other advantages. It is faster, needs comparatively little labour, and can be closely controlled; it is possible to vary the tin coating thickness within certain limits as well.

Because it needs costly equipment electrolytic tinning, like the continuous wide strip process from which it obtains its steel sheet, can only be justified where outputs are large. This is why it originated in America. But it soon spread elsewhere, Ebbw Vale being the first works outside America to start electrolytic tinning. The process is now in international use and virtually all the tinplate made in Britain is electrolytically tinned.

When the British iron and steel industry faced reconstruction and re-equipment after the war it had behind it much that was to its credit, though it had been for some years prior to the outbreak of hostilities static on the whole. This was in a large measure due to the lean years between the wars, which had given little encouragement to develop. It had a lot of leeway to make up but it was not alone; so had the steel industries of many other countries. Blast furnaces had grown in size and output during the first forty years of the twentieth century, mechanical charging and the use of pig casting machines were widespread though not universal, and the principal means of tonnage steel making was the open-hearth furnace.

Acid Bessemer steel was still made in a few places where local conditions favoured it, but the process was not of great importance. The basic Bessemer process had disappeared altogether for a time by 1925, though it had been revived again at Corby, Northamptonshire, in 1934. Here it was used as the steelmaking process at a new integrated iron, steel and tube works, and it was employed simply because it suited the iron made from the local iron ores. But although it did very useful work at Corby, the basic Bessemer process, like the acid, was never again to be of major importance in Britain.

Steady progress had been made over the years in the mechanization of rolling mills especially, of necessity, those producing heavier sections and plates, but there were still many hand mills in operation. This was particularly the case in the sheet and tinplate sections of the industry, where, at the end of the war, the major part of the output was from hand mills, often in very small works. Carbon and alloy steels had become increasingly the products of the electric furnace, but here again, the old methods were still in use. There were still numerous crucible furnaces in 1945.

The principal difference when the British steel industry began its reconstruction after the 1939–45 war was that by that time the industry had become international in the matter of techniques and processes. It was a development which was only to be expected and had, in fact, been going on for a long time. A hundred years previously the British industry had been to the forefront in technical developments; it had taught the world how to make iron and steel. But a world which learns new techniques does not stand still. It goes on to develop still further new ideas, and it does so for more than one reason. America, as we have seen, was compelled to find ways of mechanizing in order to expand, and its ideas, or at least the best of them, were recognized and adopted elsewhere, including in Britain. This process of development went on out of sheer necessity in other countries, too, and again the best of the ideas were adopted elsewhere. It should be made clear, though, that if Britain no longer held a virtual monopoly in ideas, she was far from being out of the race, as will be seen when considering the events of the last few years.

CHAPTER EIGHT

Steel from 1945 to the Present

Technical development has proceeded at a tremendous rate since the end of the 1939–45 war, and nowhere has this been more marked than in the iron and steel industry, which has not only invented new techniques for itself but has also been very ready to make use of new ideas in other fields, such as electronics. Today no single nation leads the world in steelmaking technology and although some nations naturally have much bigger outputs than others, some of the smaller nations, in terms of output tonnage, have outstanding developments to their credit. Iron and steel, from the technical point of view, are now completely international, and such is the co-operation between nations and between firms that no worthwhile development goes unnoticed by the industry on a worldwide basis. It is no longer possible to think of the steel industry of Britain, or of any nation, in isolation, though the individual contributions leading to the international whole must be considered. Britain, though no longer the world leader in iron and steel technology, has no reason to be ashamed of her own contributions; in any case, as has been said, there is no single leader in technology now.

For ironmaking the blast furnace remains supreme. In Britain it is the only means employed for this purpose as it is in most other industrial countries. But it is not the only device for reducing iron ore to iron. Direct reduction, using electricity or natural gas, is practised on a fairly small scale in some countries, where coke is scarce or non-existent and an alternative is available. Electricity is used in Switzerland, for example, and natural gas is used in Mexico. And if the blast furnace continues to hold the field in Britain, there is no certainty that it will continue to do so indefinitely. Technology changes very rapidly and what seems immutable today may well change tomorrow. The natural gas resources off the coasts of Britain may well bring about radical changes in both iron and steelmaking.

But for the present the blast furnace remains the sole British commercial means of ironmaking, and it is unchanged in principle, though it is of course vastly increased in size and output, and all the most modern examples are highly mechanized. Some, in fact, can be said with truth to be almost completely automatic in operation. The growth in individual output has meant, naturally, a drastic fall in numbers of blast furnaces, and Britain now has only 86, compared with 190 in 1939. Compared with only two decades earlier, the fall in numbers is even more startling for the total in 1927 was 427. A century ago there were 655.

Individual outputs from the furnaces of today can be quite impressive. One of the furnaces at Margam, South Wales, for instance, which has a hearth diameter of 31 ft, produces over 2,500 tons of iron a day. Again, the degree of automatic operation is remarkable. There are no men normally at the top of the furnace today, and no barrow-wheelers. Maintenance men go up the furnace from time to time; otherwise the top is entirely unmanned. The two modern furnaces at Spencer Steelworks, Monmouthshire, are fully automatic as regards charging, operating to a pre-set programme and varying the charging rate automatically to suit changing conditions. They are not alone, for the same sort of operational pattern is general, though details vary between individual plants. Today a blast furnace, once it is blown in, receives but little attention from operatives and works on at the maximum rate (given a good enough state of trade) until it needs relining and is blown out. Even the relining, which was once a leisurely operation, is aided by a variety of mechanical devices.

There have been changes in the raw materials, too, for it is no longer the general practice to charge iron ore just as it is mined, or at best screened to take the dust out of it. Several means now exist for upgrading or beneficiating* the ore. These are variously applied according to the nature and physical condition of the ore itself; the most common method of preparing the ore in Britain is by sintering. This, as has been said, was an American idea (incidentally, it is by no means confined to iron ores), but it was Britain which first made successful use of sintering to prepare the whole of the furnace burden.

---

* This is a trade term which covers the concentration of the ore, by one or more of several processes, to raise the iron content. It is distinct from processes such as sintering, which improve the physical qualities.

Sintering was originally applied to the ore fines or dust which, while chemically suitable for reduction to metallic iron, were unusable, or at least a great source of trouble in the furnace as they caused clogging in the furnace stack. By sintering or agglomerating the ore fines into lumps of open texture which the furnace gases could penetrate, the furnace could deal with the otherwise troublesome material easily. Experiments on sintering carried out at Appleby-Frodingham Steelworks, Scunthorpe, Lincolnshire, led to the use of an all-sinter burden. This, introduced in 1952, was the world's first. It has since become standard practice at Appleby-Frodingham, where it has resulted in outstanding success in the production of iron from low-grade local ores, and all-sinter burdens are now used in several parts of the world.

While the ironmaking side of the industry has so far only developed existing principles to a high degree of efficiency, the steelmakers have brought into use some completely new processes, with remarkable results. The major development in steelmaking has been the use of oxygen in bulk (or as it is called, tonnage oxygen) in place of air or oxygen-rich solid materials as used in the Bessemer converter or open-hearth furnace. It has been so successful that oxygen steelmaking in one form or another is beginning to oust all other processes for bulk steelmaking.

Actually the idea of using oxygen gas for decarburizing iron to make steel is quite old. It was patented by Bessemer subsequent to the development of his converter process, which used astmospheric air, but although he knew full well what he was proposing, he was unable to put the idea into practice because at the time oxygen was not available in even moderately large quantities. In fact bulk oxygen did not become readily available until after the 1939–45 war, when it was made by the liquid air process.

Experiments with the use of oxygen for steelmaking had been made in Germany before the war, and when the war ended further trials were made, on a small scale, in Switzerland. But it was in Austria that the process first became a real success. Austria is not a large steelmaking country, and the problems which arose there when the war ended called for an expansion of output for which conventional steelmaking methods were not suitable. Steel was needed for reconstruction, there was too little scrap available for the open-hearth furnace, and the local ores were unsuitable for the

Bessemer process. The only thing the Austrians had was plenty of iron made in the blast furnace. To use this, in molten form, they designed, in 1953, a converter vessel, not unlike the Bessemer vessel in shape, but having no tuyeres in the bottom, and they blew commercially pure oxygen at high velocity on to the top of the molten metal. The oxygen was blown through a water-cooled lance, which could be withdrawn upwards so that the vessel could be tilted for charging and tapping. A very fast reaction resulted and good steel was made in about half an hour.

The original Austrian name for the process was Linzer Düsenverfahren, and the initials LD have now been universally applied to it. By coincidence, the process was first used successfully in steelworks in Linz and Donawitz, and the initial letters of these towns are now usually said to have given the name LD to the process. In fact the first explanation is the true one, but it is a matter of little importance, especially as when the process was taken up in America it was, rather confusingly, called the Basic Oxygen Steel or BOS process, and this name is now spreading to Britain.

What really matters about the LD process, whatever local variations of the name may be, is its outstanding success. So successful has it been that every major steelmaking country now uses it, or a variant of it devised to suit high-phosphorus irons, and about 120 million tons of world steelmaking capacity is now based on the original Austrian process of 1953. Already the LD process accounts for a substantial proportion of British bulk steel output, and it was only put to use in this country in 1960; the credit for being the first works to use the process in Britain goes to Ebbw Vale. Since 1960 the LD process has spread to many other steelworks and one, Spencer Works, near Newport, Monmouthshire, uses it exclusively. Moreover, LD steelmaking is still spreading.

The first LD vessels or converters were quite small, those at Linz in 1953 having a capacity of only 35 tons. Today the vessels are very much bigger, and they are getting bigger still. Those at Spencer Works are of 145 tons capacity, and there are others of the same tonnage at Consett, Co. Durham. Vessels of 270 tons capacity are the basis of a current development scheme at Abbey Steelworks, Port Talbot, Glamorganshire. Abroad 300 ton vessels are already in use, and there is every reason to believe that the limit of size is not yet in sight.

Blowing with oxygen takes thirty to forty minutes and allowing for charging with molten metal and scrap, slagging, tapping and incidental operations, it is theoretically possible for a vessel to make its rated output hourly; thus, the 270-ton vessels could produce, in theory, 270 tons of steel an hour or, in continuous operation, 6,480 tons a day. Of course, in practice the production is less than the theoretical figure, but it is still very great, and it needs no more labour than the older processes. On the debit side, the plant is large and costly, and, since it produces while blowing with oxygen, very large quantities of dense brown or orange fumes, expensive equipment is needed to clean the gases before they are released to the atmosphere. At least this is so in Britain, under the provisions of the Clean Air Act and most other industrial countries have some similar requirements. But the LD process is so fast that it is easy to see why it has spread so widely.

There are two other oxygen steelmaking processes, the Swedish Kaldo and the German Rotor, and both are in use in Britain, though on a much smaller scale than the LD. The new steelworks built at Shelton, Staffordshire, in 1964, uses the Kaldo process exclusively. Oxygen is also used to speed up refining in the open-hearth and electric furnaces, but its major application is in the oxygen-blown converters, particularly the LD, and the pattern which has already emerged for future steelmaking is that the bulk of tonnage steel requirements will be met by the oxygen process,* while the open-hearth furnace will gradually become extinct. It is certainly unlikely that any more open-hearth furnaces will be built.

Making steel so fast brings other problems with it. Movement of molten steel from the relatively small producing area to the point

* At the time of writing a revolutionary new process is undergoing commercial evaluation; it has already been proved to be practicable on a production scale. This is the spray steelmaking process, developed by The British Iron and Steel Research Association. Molten iron from the blast furnace is allowed to fall at a controlled rate through a ring of oxygen jets and powdered lime is blown into the stream at the same time. The equipment required is comparatively simple and the outstanding feature of the process is that it is continuous. The oxygen and lime provide the steelmaking reactions and as long as they and the stream of molten iron are kept up at the correct rates, it is literally a case of 'iron in at the top, steel out at the bottom'. Every other steelmaking process hitherto devised is on the batch principle. The modern ones are fast but they still operate in batches. Spray steelmaking, too, is very easy on refractories, which some of the others are not.

at which it is cast into ingots needs careful organizing, for example. But the biggest problem is that of analysis. When steel is made in an open-hearth furnace there is time to take samples and analyse them so that the correct time for tapping can be determined accurately. With oxygen steelmaking the steel is made so fast that the older, laboratory methods of analysis are too slow; an analysis could be out of date before it was finished. Moreover, specifications are nowadays much stricter than they used to be, so aggravating the position.

Very fast methods of analysis have therefore had to be devised, and the most impressive of these is the automatic spectrographic process. In this a sample of metal is taken, cooled, cut in two, polished on the cut surface and put into the automatic spectrograph. An electric arc is then struck between an electrode and the sample, the light given off is read spectrographically by electronic instruments and the percentages of a dozen or so elements are thus determined. These percentages are then typed out automatically in the laboratory and in the melting shop. All this is done in not more than five or six minutes, and it is likely to be speeded up; speed and accuracy are now vital. Automatic spectrographic analysis is now being used in electric furnace plants as well, and, incidentally, for dealing with non-ferrous metals. The equipment is complicated and expensive, but wholly justified by the results.

Innovation has not been confined to the making of steel, it is equally apparent in the processing of it, too. Continuous casting is an outstanding example of what has been done in this field. To understand continuous casting, however, it is necessary to take a look at the conventional means of dealing with the molten steel from the steelmaking furnaces.

Traditionally, steel which is to be rolled has always been poured or teemed into heavy cast iron ingot moulds, which shape it into heavy rectangular or square (and sometimes octagonal) ingots. The moulds are drawn off (or stripped from) the ingot as soon as it is sufficiently solid, and the ingot then goes, usually, to a soaking pit, where it is brought up to the required heat for rolling. Sometimes the ingots are left to go cold and then reheated; sometimes they are charged hot to the soaking pit where the heat soaks uniformly through them. In either case the ingot goes, when heated to the correct temperature, to the first or primary rolling mill, where it is given the necessary number of passes to reduce it to a flat slab (for

Hertfordshire Libraries
Central Resources Library
Kiosk 2

Borrowed Items 28/09/2012 12:31:10
XXXXX7530

Item Title	Due Date
English medieval industrie	19/10/2012
Between the woods and th	19/10/2012
Mass extinctions	19/10/2012
* Sedimentary rocks in the	19/10/2012
* PHOENIX AT COVENT	19/10/2012

* Indicates items borrowed today
Please remember to unlock
your DVDs and CDs

Enquiries and Renewals phone number
0300 123 4049
or go to : www.hertsdirect.org

plate and sheet rolling), or to a square bloom (for further reduction into any other products such as joists or channels, or billets for rolling into smaller sections, rounds, squares etc.). Rolling to slabs (slabbing) and rolling to blooms (blooming) are usually done in separate mills designed for the particular purpose, but some mills can roll both types. This has been the standard procedure since the second half of the nineteenth century when steel, as distinct from iron, began to be rolled on a large scale. It remains the practice in many works, but continuous casting is now well established and spreading.

In continuous casting there is no primary mill. Molten steel is cast direct from a ladle into the form of a slab or bloom, which issues continuously until the ladle is empty and is cut off automatically into the required lengths for secondary rolling. The process has the economic advantage that it dispenses with some very costly machinery, but this is not all. Production by either process results in a certain amount of scrap (which can be remelted but has only scrap value) because the ends of the products have to be cut off and trimmed. Continuous casting produces much less scrap than primary rolling or, to use the steelmaker's term, the yield is better.

Continuous casting has only come into successful use in recent years, but the idea is actually more than a century old. Bessemer was the first to suggest that molten steel could be cast continuously, but his idea was not a success, and neither were the ideas of others who followed him at first. The process was first used commercially for non-ferrous metal, which need not concern us here, and it was not until after the 1939–45 war that it was applied with any success to steel. Today the process is broadly based on that developed by S. Junghans in Germany, with contributions from Britain, Russia and other countries.

It is fundamentally simple but in practice calls for extremely careful control. Liquid steel, at a temperature of about 1,600°C is poured at a controlled rate from a bottom pouring ladle, through a refractory-lined tundish into an open copper mould of the size, in cross-section, of the required bloom, billet or slab. This mould is hollow-walled, and cooling water is circulated between the walls, so that the outer skin of the molten steel solidifies rapidly. The mould reciprocates vertically over a short stroke, which assists in the prevention of sticking of the molten metal to the mould walls. At the bottom of the mould the cast product has solidified sufficiently for it to be drawn out downwards continuously at the same rate as the molten metal enters at the top. Water sprays immediately below the mould cool the cast product further, and it then passes between power-driven rolls which grip its surface and provide the downwards drawing motion.

To start the process a dummy bar is entered into the mould from below and the first stream of steel entering the mould solidifies on the top end of the dummy bar. This is drawn downwards, carrying the cast shape with it. The dummy bar is then removed, for use at the next cast, and the cast product is drawn out by the withdrawal rolls until the ladle of steel is exhausted. As the cast bar moves on downwards it is cut into lengths by a pre-set automatic flame-cutting machine and the lengths are taken away for further processing.

Britain was quite early in the field with continuous casting, a small machine having been put into operation at Bradford in 1946. Another machine came into commission at Barrow-in-Furness, Lancashire, in 1952 and this machine, though installed for commercial purposes, has also been used for experimental work. Much valuable operating information has been obtained from the Barrow machine. Shelton Iron and Steelworks, Stoke on Trent, installed continuous casting machines in 1964 to take its whole output of steel and so became the first works in the world to rely entirely on the process.

The latest developments in continuous casting have been the reduction, in two stages, of the overall height of the machine. To cast the metal into the mould, withdraw it, cool it and cut it to length, all vertically, needed a machine sometimes 100 ft high and this was often a disadvantage from the operating point of view, besides requiring costly structural steelwork and buildings. In about 1961 the arrangement was altered so that the cast bar, immediately below the withdrawal rolls, passed through a further set of rolls which turned it at right angles so that it issued parallel with the ground, and was cut to length there. In 1964 an international company, with British participation, introduced a continuous casting machine in which the metal is cast in a curved mould, cooled and withdrawn in the same curve, and sent out at ground level where it is automatically straightened and then cut to length. This has reduced the overall height of the machine to about a quarter of what it was and the curved machine has been widely, though not exclusively, adopted. At the end of 1965 there were about 120 continuous casting machines in use in various parts of the world, and they are spreading very quickly.

Progress has extended to the rolling of steel, too. Speed of operation of mills and accuracy and consistency of product, have been improved greatly by new techniques. In particular this applies to automatic control of thickness of flat products such as wide and narrow strip. Control of thickness or gauge is never easy, but when sheets were rolled individually by hand there was time for mill adjustments between the rolling of individual sheets. With continuous strip mills producing strip at speeds up to about 60 mph, rolling off-gauge, which could happen very easily, could be extremely serious, for it could mean that a lot of scrap material was produced. Automatic gauge control has proved the answer to this problem; the best known system is that developed by The British Iron and Steel Research Association and licensed for manufacture in many countries. Automatic gauge control (generally known as AGC) usually works by continuously measuring or monitoring the thickness of the strip as it leaves the mill and sending back automatically corrections to the mill settings in accordance with the readings obtained from the thickness-measuring instrument. This control in the last twelve months has been taken even further by another British invention. In this, the Loewy-Robertson 'Constant-Gap' mill, the rolling forces are measured while the strip is being rolled, and corrections are fed continuously into the mill to allow for all the factors which can cause gauge variation. The control is fully automatic, and once the required gauge has been set, the mill maintains it to a remarkable degree of accuracy; consistent accuracies of ± 0.0002 in have, in fact, been obtained in practice.

High speed automatic rolling has also extended to small bars and rounds and the term 'high speed' takes on its full significance in mills such as one produced recently in Britain for rolling small rounds at finishing speeds up to 10,000 ft/min (about 120 mph). At such speeds, of course, the finished product could not possibly be handled manually, and it is, in fact, coiled in much the same way as strip. Mills running at very high speeds, whatever they are rolling, need very precise control of the individual drives, and it is only the advent of very sophisticated electrical and electronic gear which has made such control possible. Today the electrical engineer shares with the mechanical engineer and the metallurgist a large part of the responsibility for successful operation of a steelworks.

It is a combination of the work of these three professions which has brought automation, in its true sense, to steelmaking. Automation, it should be pointed out, is not simply automatic operation, though the two are all too often confused. Automation is a vogue word, applied indiscriminately to any process which works without human aid. But in fact automation must contain at least an element of self-inspection combined with automatic feed-back of corrections to the machinery if anything goes wrong. Automatic gauge control, in which the equipment measures the material gauge continuously and sends back correcting signals to the mill if the gauge is not within the specified limits is true automation. Automatic operation of a mill according to a programme set into the controls, though it may look very complicated, is not automation, unless it inspects its products and feeds back corrective signals. In fact, automation need not be complicated at all, though it often is.

A more interesting development, in some respects, has been the introduction of the computer to take over some routine tasks formerly carried out by mill operators. This again is often thought to be automation which, if it does not fulfil the feed-back requirement, it is not. Nevertheless, it involves the use of some advanced equipment. An important feature of the computer is that it can not only be used to relieve men of tedious routine work, but that it can also do it at a tremendous speed. An early British example of this is the computer-controlled billet cut-up line installed at Stocksbridge Steelworks, near Sheffield, in 1961. Very high grade alloy steels are made at Stocksbridge, and since these are expensive, the best possible use must be made of every billet rolled. This would be fairly simple if every billet came out to an exact length and every order was for this length or an exact part of it. But in practice, the length of a billet cannot be guaranteed as it comes from the mill; it varies for one or more of several reasons, including the fact that in order to ensure top quality, more or less material may have had to be dressed off the bloom from which the billet was rolled. Customers' orders naturally vary, too. Conventional practice was always to measure the length of each billet and a skilled man would then decide how it could best be cut into lengths, to meet the orders and give the minimum of waste.

At Stocksbridge the length of the billet is measured automatically and the figure obtained is fed into a computer together with a statement of grade and quality. Information about orders on hand, lengths, qualities and grades required is also given to the computer, which works out the most economic way of cutting the billet, allowing for the small loss in cutting, and gives the result digitally (that is in a row of figures like a cash register) to the saw operator. The computer gives this information to the man at the same time as he gets the billet; no time is lost and no mistakes are made. There is no theoretical reason why the computer, having worked out the cutting lengths, should not set the saw automatically, and this type of development is already in hand. In fact, fully automatic operation of more than one type of rolling mill is already an accomplished fact; once the programme is set the mill will do what it has been instructed to do indefinitely. But this does not mean that men will no longer be required. Far from it. The men will, in future, find themselves increasingly relieved of tedious mental tasks, just as the mechanical developments referred to earlier have taken away many (in modern works most) of the hard, dirty, unpleasant and even dangerous jobs which had once to be done manually.

In the alloy-steel field progress in the last few years has been equally impressive, but automatic operation especially in rolling, will not be found so much here since the quantities of alloy steel required are much smaller than those of mild and carbon steels, and a very high degree of mechanization or automation can only be justified by a very large output. It is in the matter of actual steelmaking, especially with regard to quality control and improvement that alloy steels have developed particularly.

Crucible steelmaking is now extinct, and alloy steels are made in electric furnaces, both arc and induction. Some of the arc furnaces are very large. There are two at Tinsley Park, Sheffield, for example, each having a nominal melting capacity of 100 tons. The induction furnace has spread particularly for the making of the very complicated alloys, which are needed in relatively small quantities, and these furnaces are much smaller; up to 5 tons capacity is fairly common, but furnaces of only a few hundredweights capacity can also be found. Both arc and induction furnaces have the great advantage of being very clean in operation, since there is no fuel there is no possibility of contamination of the metal by sulphur or any other element in the fuel. This is of vital importance today when the standards of quality called for by users of alloy steels are so high. In many cases these users cannot tolerate even traces of any foreign element, and some of the standards of today are so exacting that even electric-furnace melted steels will not meet them; further refinement is necessary. So vacuum treatment has come to the alloy steel industry. This was first used out of necessity for dealing with some of the newer metals such as titanium, and has spread to steelmaking. Some of these newer metals cannot be melted at all in contact with air, and means had to be found for putting an induction furnace inside an air-tight chamber which was sealed and evacuated before melting started, and held under a vacuum for the whole of the melting period.

Vacuum melting was equally applicable to steelmaking and it has been found useful for steels to very close specifications; a typical example is a steel for gas-turbine parts. Steel is made carefully in electric furnaces in contact with the air in the normal way and then remelted in a vacuum furnace. None of the gases normally present in the air and which could contaminate the alloy, can get to the molten steel, and any gases present in the steel and liberated on melting are drawn away by the vacuum pumps, leaving a very clean steel in the furnace. Vacuum melting is essentially a melting and refining process; it is not a steelmaking process, the steel being first made elsewhere.

Refinements in the equipment, however, do enable changes to be made in the steel composition. Remotely controlled chutes and vacuum locks make it possible to add alloying ingredients to the charge of steel, while it is still under vacuum, and some vacuum furnaces have a secondary chamber, also evacuated, in which the steel, when ready, can be cast into small ingots or into moulds, which can then be held under vacuum until they are sufficiently cool to suffer no ill-effects from the air. None of these vacuum furnaces is very large compared with other steelmaking furnaces, up to about a ton capacity being typical. They are essential for some of the very high grade alloys which are only needed in small quantities, and they meet this need fully. There is no reason why they should be made larger (except on the grounds of expense), but there is not at present any call for very large quantities of vacuum remelted steel.

Vacuum remelting can also be done by what is called the consumable electrode process. In this process steel is also made to the required composition in a furnace in contact with the air, but in this case it is cast into a large round bar or electrode weighing a ton or more. This is then inserted vertically into a vacuum chamber and an electric arc is struck between it and a small pool of molten metal in a water-cooled copper mould. As the electrode melts, it falls in droplets into the mould and there builds up to form a solid ingot, the little pool of molten metal remaining on the top. The electrode is gradually melted away and all the metal, in molten form, is subjected to the vacuum, which draws off unwanted gases. This process is purely one of remelting. It uses simpler plant than the vacuum induction process, and can form quite a large ingot.

Vacuum remelting is naturally expensive, since it requires costly plant, vacuum pumps and control gear, but it is justified by results on the more expensive alloys. More recently a different remelting process has been introduced which has some points in common with vacuum remelting, but does not, in fact, take place in a vacuum. It therefore uses simpler equipment. This is the electro-slag refining process. As in vacuum arc remelting, an electrode of suitable composition is first made by conventional means. This is then entered vertically downwards inside a water-cooled copper mould, and an arc is struck between its end and a piece of steel in a quantity of slag, which soon melts. The end of the electrode is then lowered into the slag, and melting proceeds by the resistance of the slag to the passage of the electric current. As the end of the electrode melts off droplets of metal pass through the molten slag to solidify in the mould, and there is intimate contact between slag and metal; impurities pass into the slag and the metal is left clean. Because there is only a small quantity of metal molten at a time the solidification takes place uniformly in a way which is particularly beneficial to some alloy steels; tungsten high-speed tool steels, for example, are greatly improved by electro-slag refining. This process is complementary to, not competitive with, vacuum remelting. Each will do some things better than the other, and each is particularly suited to certain types of work. Both are of increasing importance.

Vacuum treatment is not now confined to alloy steels. It is also applied increasingly to bulk steels, though the methods employed are different. For bulk steels the normal procedure is to subject the molten steel, after it has been made in the conventional manner, and before it is cast into ingots, to a degassing process. This can have one of several objects, according to the analysis of the steel and the use to which it is to be put. A typical application is for removing gases, particularly hydrogen, from steel to be used for making large forgings, such as steam boiler drums. The presence of gases in such steels has always been a source of trouble, causing cracking and embrittlement.

Hitherto the only way to get rid of these gases was to subject the steel to a very lengthy process of heat treatment. This might take weeks and could be expensive. Vacuum degassing is just as effective and much quicker; it may take only a few minutes in fact, and never takes longer than about half an hour.

There are various ways in which molten steel can be degassed. The simplest is the chamber process. In this process a ladle containing 100 tons or more of molten steel, which has been made by conventional means, is lowered into a large steel chamber, which is then closed and clamped airtight mechanically. Very powerful vacuum pumps are then set in operation, and the chamber is evacuated rapidly. The vacuum is held for anything from ten to twenty-five minutes, and unwanted gases are drawn off the molten steel. Sometimes, to ensure that all the steel is subjected to the vacuum, it is stirred by means of an electric induction coil in the chamber. At the end of degassing the vacuum is broken, the ladle is lifted from the vessel, and the steel is poured or teemed in the normal way.

Other methods of degassing include progressive degassing, and stream degassing. In the first of these an evacuated vessel has a pipe at the bottom which dips into the ladle of steel to be degassed, a proportion of which is sucked up into the vessel, where degassing takes place. The vessel is then raised and most of the metal is ejected back into the ladle. By repeating this lifting and lowering twenty or thirty times, the whole contents of the ladle are passed through the degassing vessel.

A variant of this process is circulation degassing. In this the degassing vessel has two pipes at the bottom, both of which are dipped into the molten metal. An inert gas, argon, is then introduced into one pipe, and this, passing upwards into the evacuated vessel, causes the molten steel to ascend with it. The steel then falls down the other pipe, back into the ladle.

Stream degassing involves the passage of molten steel from a ladle into a second ladle, of the same size, which is held in an evacuated

24 Pouring molten steel at a continuous casting plant, Shelton, Stoke-on-Trent

25 Discharge section of continuous casting plant, Shelton, Stoke-on-Trent. The cast billets emerge from the curved guides in a horizontal position

26 Outgoing side of roughing stand of universal beam mill at the Appleby-Frodingham Steelworks, Scunthorpe

27 Outgoing side of finishing stand of universal beam mill at the Appleby-Frodingham Steelworks, Scunthorpe

vessel. The steel, falling in a relatively thin stream, breaks up into droplets in the vacuum vessel, thus exposing a large surface to the degassing action. This process itself is sometimes modified, an ingot mould being substituted for the second ladle. The steel is then degassed as it is being cast into an ingot. Finally, there is The British Iron and Steel Research Association's continuous process, in which the steel, by a siphon action, is caused to pass from a tundish (which is kept filled from a ladle) into a vacuated vessel and out again in a continuous stream.

Degassing, by one process or another, is now in widespread use, and is likely to increase. Its use is all part of a development which has been going on for a long time, and has accelerated enormously in the last few years—the combination of economic production, increased output, and the meeting of tight technical specifications which are always getting stricter. To all this is added the constant search for and development of, new types of steel and iron. There is no end in sight, and all those who have the welfare of the industry at heart hope there never will be. For the end of development would surely mark the beginning of the end of all those many products which can be grouped together in the one word iron.

CHAPTER NINE

Ironfounders and Steelfounders

It is logical to devote separate chapters to consideration of the production of iron and steel articles by casting or forging since both, though for long associated with ironmaking and at one time integral with it, gradually became the separate and distinct practices they are today. Of the two, smithing, or forging, is by far the older. The primitive ironmakers were also the smiths; they made the iron and they shaped it by hammering it into the tools and weapons which were the only forms in which it was used. This pattern would persist as long as production was on a very small scale. As outputs increased, so a division of labour became essential, and ironmaking and ironworking developed into separate trades, though they may well sometimes have been under the same employer and in the same works.

Ironfounding did not develop until after the blast furnace came into use. It could not, for it needs molten iron, and there was no such thing (except occasionally by accident), before the blast furnace. Even after the blast furnace made molten iron available, ironfounding took time to develop, for there remained the question of what articles could usefully be made of cast iron. Casting itself, the art of making a mould into which molten metal is poured, to cool and set in the shape required, antedates ironfounding by centuries. It was known in prehistoric times that metal objects could be made in this way, and early castings of bronze, gold and other metals, both for use and for ornament, can be seen today in museums.

But cast iron, as has been pointed out, is hard and brittle; in thin sections it can be broken quite easily. It would never serve for, say, a knife, an axe or a nail, which all had to be forged from wrought iron. In time, as we shall see, developments in the metallurgy of cast iron itself made it possible to use the metal for many things for which it had hitherto been wholly unsuitable, but the first of these developments came only in the early eighteenth century and some of them date only from the last twenty years or so. At first cast iron was simply what came from the furnace and it had to be accepted or rejected solely on this basis.

So at first cast iron was used for very simple articles; firebacks, which can still be seen in old houses and in museums, are a typical example. Grave slabs were similar in design and were also made of cast iron. These were quite easy to make. Anyone who has pressed his fingers into damp sand on the seashore has made a rudimentary mould. If a piece of wood had a design carved on the surface and this surface were pressed into a level bed of slightly damp sand, the design would be left in the sand when the wooden master or pattern was removed. Molten iron poured into the depression thus left would solidify and take on the design of the pattern. This process could be repeated, for the pattern was not affected; only the sand mould was destroyed as the casting was lifted out of it. So any number of castings could be made from a single pattern. Naturally, in time, the wood pattern showed signs of wear, especially if it was handled at all roughly, and it had to be repaired. But the fact that a single pattern would serve to produce a large number of castings, all identical, was an important distinguishing feature of casting. Forgings did not have this identical repetitive quality; each one had to be made individually by the appropriate hammer work.

An important development took place in 1543, when the first cast iron cannon to be made in Britain was cast at Buxted, Sussex. Shot for cannons was also produced in cast iron and in the seventeenth century a few other odds and ends, such as garden rollers were made of the same material.

All these early castings were of simple shape and with the exception of cannons, they were solid. Cannons, of course, had to have a hole up the centre. But the best way to make a hollow casting is to 'core' out the hollow portion and this is the method used today. Coring can easily be understood by taking a simple example of a hollow casting such as an ordinary cast iron pipe, and seeing how it could be made. The words 'could be' are necessary, as cast iron pipes can be, and today are, also made by a different process which involves no core at all; this, however, will be discussed at the appropriate point in the narrative.

If a wooden pattern of the size and shape of the outside of the pipe required is surrounded by sand and then withdrawn, the mould, if filled with molten iron, will produce a solid casting, not a pipe. It should be noted, in passing, that metal occupies more space when it is molten than when it is solid, and allowance has to be made for this fact in the pattern, which has to be slightly bigger than the finished article is required to be. For ordinary cast iron this allowance is about $\frac{1}{8}$ in for each foot; it varies according to the casting and again for other metals.

If a cylinder of sand, of the diameter of the required hole in the pipe is supported in the centre of the mould after the pattern has been removed, this leaves only a tubular space for the molten metal, which is held outside by the sand of the mould, and inside by the sand of the core. This is the basis of hollow casting, but in practice it is much more complicated. For example, if our specimen pipe were, say 6 ft long, 3 in outside diameter and $2\frac{1}{2}$ in inside diameter, leaving the wall $\frac{1}{4}$ in thick, it would need a core 6 ft long and $2\frac{1}{2}$ in diameter, which could not be made self-supporting in sand. So it is necessary to build the sand up on an iron bar. Moreover, the molten metal will release gases from the sand which have to be allowed to escape freely through the sand or they will penetrate the molten metal and form blow holes. This can be done from the outside of the casting by pushing a wire through the sand to the pattern to make a lot of little holes, too small for the metal to penetrate, but large enough for gas. In the core other arrangements have to be made, for molten metal surrounds it completely. Venting in this case can be done in more than one way, a common one being to interpose a layer of straw between the metal core bar and the sand of the core.

There is another important matter to be considered when a casting, solid or cored, has any shape other than the flat fireback or slab. Flat shapes can be pressed into the sand and lifted out again. Our hypothetical pipe and most other things which are cast could not. So it is necessary to divide the mould into at least two parts, which can be taken apart to remove the pattern and put together again in exactly the same relationship, leaving the cavity required. This dividing of the mould can be quite complicated with some castings, but our cast iron pipe will serve to show its fundamentals.

Such a pipe could be moulded in a two-part moulding box by first filling one half of the box with damp sand and then setting the pattern in it until its longitudinal centreline is level with the top of the box. The 'box' is, in fact, a pair of open frames, with neither top nor bottom. Some dry powder (sand or flour, for example) is then sprinkled over the flat surface of the sand, the second part of the box is placed in position on the lugs and pegs which locate it, and more damp sand is placed in the box and rammed firmly into position. The top part of the box (the cope) is then lifted off the bottom part (the drag) and placed on one side while the pattern is removed. The dry parting material has prevented the sand in the two halves of the box from sticking together. It now only remains to cut passages (a runner and gates) through the sand with hand tools to allow the metal to enter the mould, place in position the core, which has been made in the meantime, and replace the cope on the drag. It returns to exactly the correct position, being located by the lugs and pegs.

The mould can now be poured. When it is cool enough, the sand is knocked out, the casting removed, and the sand core, in turn, is knocked out of the casting, which, when the runner and gates have been cut off and any rough edges removed, is ready for use. The moulding box is ready for re-use as soon as the sand is knocked out, and the pattern and the core box in which the core is made can likewise be used indefinitely. This is only a very simple example of coring. Many castings are cored to much more complicated shapes. An older example is the valve chest of a steam engine and a modern one the cylinder block of a motor car engine. Both these call for complicated coring. But simple or complicated, the principle of making hollow castings remains the same.

Abraham Darby I did very useful work in the development of iron foundry practice (besides his more famous activities in coke smelting) for he was primarily a maker of cast iron pots and other hollow-ware, and it was after he had perfected his moulding process that he turned to the question of his iron supplies. Darby patented his moulding process in 1707, and although his patent does not give any technical details other than a statement of the moulding material, it is possible to work out what was involved. The patent covered the making of what became known as 'bellied' pots, that is pots which swelled out to a greater diameter in the centre than at the top and bottom. They served as an all-purpose cooking and water-heating pot, having three short legs and a half-loop or bail handle by which they could be suspended over a fire, and they were made in many sizes right up to recent years. For fairly obvious reasons the larger vessels were facetiously called 'missionary' pots, and the name stuck, though the alternative name 'Kaffir' pot was generally used in correspondence and catalogues. The important feature about these pots is that the pattern would have to be made in two pieces to get it out of the mould, and the moulding box would need to be in three or four parts. It represented, in fact, quite an advanced piece of moulding. Darby's patent specifies that the moulds were made in sand only, without loam or clay, and it is necessary at this point to consider the nature of the moulding material itself.

So far reference has only been made to 'sand', and in a general sense this is sufficient, but in fact there were three principal moulding materials, loam, dry sand and green sand, which appeared on the foundry scene in that order chronologically. When the earliest castings were made in the pig bed the sand of that bed was good enough; any naturally occurring sand was suitable if it were reasonably resistant to heat. For pigs it could be coarse as well, but for, say, a fire back with an intricate pattern, a little fine sand would be placed in the moulding position before the pattern was pressed in. This gave rise to the practice, followed today, of using a good quality or facing sand in that part of a mould with which the molten metal would come into contact, and old or poorer quality in bulk as a backing sand.

When castings began to be made away from the pig bed, the first moulding material was not simple sand at all but loam. This was basically a mixture of sand, clay and horse manure, though sometimes more than one sand, fire-clay, and sawdust formed part of the mixture. When thoroughly mixed with water, loam was a smooth claylike substance which could be moulded into the shape required, and patterns, in the normal sense of the word, were not used. Instead, the rough shape of the casting required was built up with bricks laid in loam or some other solid foundation was made; loam was plastered on the pile, and the outside shape was formed with the aid of a shaped wooden board. This built-up mould of loam was then dried by building a coal or coke fire near, under, or, when the shape of the mould suited, inside it. The mould was then ready for casting.

A brewing cistern, which is mentioned by Dud Dudley, will serve as a typical example of how an article could be made in loam. The cistern would be in the form of a pot, probably hemispherical and would require both a mould and a core. To make the mould a hollow shape of roughly the size of the outside of the required pot would be built up in bricks, and plastered all over with loam. An iron post was then set up in the middle of the mould and a shaped wooden former or strickle was hinged to it. By swinging this strickle through 360 degrees manually, the soft loam surface was swept (struck or strickled) to a smooth surface of the shape and size required. This was then dried. Meanwhile, the core had been made in the same way except that it was struck or strickled externally. When the core also was dried, the mould was set in position over it, leaving a space into which molten iron was poured to make the pot or cistern. Taking out the casting when it was cool enough would destroy the swept surface of the mould, but it could be built up again as required.

Loam moulding is rare today, but the next material to appear in date order, dry sand, is still used very widely, both for cores and for complete moulds; some are built up of several dry sand pieces and produce castings of very complicated shape. Dry sand has much in common with loam, in that it is dried before the mould is used, and it will contain some clay which will bind it together when it is dried. Sometimes small quantities of other binding materials are added, especially when the dry sand is used for cores; these extra materials can include flour, resin and molasses. Dry sand moulding, using patterns and moulding boxes, was certainly practised by about 1630 to 1635, and it has held its own for high class work of all sizes up to the present. Newer processes, including some for the chemical setting of the sand have now made heavy inroads into conventional dry sand practice.

The third, and by far the youngest of the moulding materials is green sand. 'Green' has nothing to do with the colour of the sand which, when new, is usually rather a dark red, but refers to its soft, undried or otherwise unhardened state, just as an unfired brick is called 'green'. Exactly when the practice arose of using moist or green sand in a moulding box came into use is not known, but it was certainly before 1785, when there is an oblique reference to it; a patent granted to Isaac Wilkinson (John Wilkinson's father) makes a distinction between sand which was dried (dry sand) and sand which was not (green sand). The earliest known reference to the use of the words green sand themselves is in the Boulton and Watt correspondence preserved in Birmingham Reference Library; it was in 1796, when the applicant for a job says that one of his sons is a 'green sand moulder'.

Green sand became the most commonly used of the three moulding materials, but it was not unusual to find a green sand mould with a dry sand core, or a loam mould with some sections added in green sand. Sometimes all three materials were used to build up a complicated mould. Green sand is very widely used today, though here again chemical methods of hardening the sand after moulding are often employed.

As has been said, the earliest castings were made straight from the blast furnace, either by making a channel to the mould from the blast furnace runner or, later, by dipping a clay-lined iron ladle into the stream of molten iron, scooping up some metal, and carrying it to the mould. As castings got bigger this became inconvenient, and the practice grew of using a separate furnace in which cold pig iron was remelted. Scrap castings, and runners and gates from castings, would also be remelted. The first separate furnace to be used for remelting iron was the air furnace. This was similar in shape to the puddling furnace (in which, in fact, the iron is first melted before it is decarburized), flames from a coal fire being reverberated down to the iron. The reverberatory furnace, as has already been stated, was brought into the iron trade from outside. Cort used it for puddling and the ironfounders used it for simple remelting. Air furnaces continued to be used extensively right through the nineteenth century, and a few are in use today for special purposes such as melting iron for casting rolling mill rolls. These are oil fired and are of a much more advanced design, but the principle remains the same.

The air furnace was a simple and effective device, but it was rather slow in operation and much better suited to melting fairly large quantities of iron at a time (modern ones melt 20 or 30 tons), than the small amounts needed by foundries making small castings. John Wilkinson provided the answer to this problem when he patented, in 1794, what is now known as the cupola. Incidentally, although the patent was in John Wilkinson's name, it is generally accepted that his brother William was the actual inventor. Very similar in form to a blast furnace (by which name it was called in the patent specification) the cupola was nevertheless fundamentally different in its purpose; it was considerably smaller, too, than the blast furnaces of the time. It was a vertical, shaft-type furnace, fired on coke and blown through three or four tuyeres, but its sole purpose was to remelt cold pig iron, which was tapped out at the bottom at intervals and cast into moulds as required. There is no real metallurgical change in the iron melted in a cupola, though in fact it does tend to pick up a little carbon from the coke. For many foundry purposes this is unimportant, though some metal specifications demand remedial measures. Cupolas at first only melted a few hundredweight of iron an hour, and they were very handy for small foundries and for the odd jobs. Today they are usually mechanically charged and there are cupolas which will melt up to 30 tons/hr or more. The cupola remains a very important item of foundry equipment, but its position is being challenged by the electric induction furnace.

As time went on iron castings found many uses, and the development of the steam engine, and of the countless types of machinery which steam power brought into common use extended the demand enormously. But the serious limitation of cast iron—its weakness in tension and bending—remained, and prevented its use for some purposes where the possibility of producing any number of identical components would otherwise have been welcome. In 1722 the French scientist Réaumur discovered a method of treating iron castings to make them malleable. He heated the castings for several days at red heat in contact with powered iron ore, and so altered the metallurgical structure that the castings emerged tough and ductile. Réaumur's process took time to penetrate to Britain, and it was not until 1804 that Samuel Lucas, a Derbyshire ironfounder, patented a similar process. Even then, progress was slow, for the material at first had a rather doubtful reputation.

In 1826, Seth Boyden, an American, developed another process for producing malleable castings, and this, though producing a similar result as far as toughness and ductility went, was different metallurgically. Boyden's castings, when broken, had a dark fracture, in contrast to those made by the older process, which was bright and 'steely' in appearance. From these differences in appearance of fracture, Boyden's process became known as blackheart, while the older one got the name of whiteheart. Both processes are still in use. There are British Standards for both, requiring a tensile strength of from 18 to 24 tons/in^2 according to grade (as compared with ordinary cast iron, which has a tensile strength of no more than about 8 tons/in^2). In recent years a third type of malleable iron has been developed; this is pearlitic malleable. This again has a different metallurgical structure, and it has even greater strength than the older malleable irons: it can exceed 40 tons/in^2. The uses of malleable cast iron are many and varied, but three examples will suffice to show its modern applications. Many of the parts of agricultural machinery, the door hinges of many motor-cars, and the supports for third-rail electrification on railways are malleable iron castings. It should be noted that malleable *cast* iron is the only true malleable iron. The term was often applied to wrought iron (in one case by no less a person than Bessemer) but such usage is quite wrong.

It only remains to complete the story of cast iron by referring to the remarkable alloy irons, which have been introduced in more recent years. These high-duty cast irons, which include Meehanite and spheroidal graphite (SG) iron (both made by proprietary processes) are produced by treating cast iron while it is molten. Thus, SG iron has magnesium added to it (usually in the form of a nickel-magnesium alloy). This modifies the structure, and when the iron cools, the carbon present is found in the form of very small nodules instead of as tiny graphite flakes, which are the source of weakness in ordinary cast iron. High-duty cast irons can have tensile strengths up to 50 or 60 tons/in^2, and are applied to hitherto unheard of uses; motor-car crankshafts are an example.

So far only cast iron (now available in many forms) has been mentioned as a material for castings. But steel can be cast as well, and for some purposes its superior strength and toughness are highly desirable. The manufacture of steel castings started in Europe in about 1845, and in 1855 Naylor, Vickers & Company, of Sheffield, introduced it to Britain. At that time a major limitation in the making of steel castings was the problem of melting the steel; only the crucible furnace existed for this purpose. This was no trouble as long as the castings were small, but when they were big the melting, carrying and pouring of the metal called for expert organization and great manual effort. In spite of this some quite large castings were made, usually as ingots for forging into such things as ships' propellor shafts. Thus, in 1859 an ingot weighing 25 tons was cast at Sheffield. As each crucible only held 100 lb of molten steel (it could not hold more, since it had to be lifted out of the furnace and poured manually), 576 crucibles were needed. The ingot was cast in five minutes, or at a rate of one crucible every half-second.

Obviously such a process was beyond the powers of all but the largest works. Not many firms could melt steel in nearly 600 crucibles simultaneously. The answer lay in the use of a different and more convenient method of melting the steel, and in 1862 the Stanners Closes Steel Company, at Wolsingham, Co Durham, put down a Siemens open-hearth furnace solely for making steel for castings; it was the first open-hearth furnace to be used for this purpose.

Mechanization was relatively slow to come to the foundry, but moulding machines, which formed the sand mould round a wood or metal pattern, developed with the demand for large numbers of identical castings. There was little inducement to mechanize moulding as long as the castings were usually different, as were the cylinders of steam engines, for example; these would most likely be of different sizes and in any case not many were made at a time. But such things as gear wheels and pulleys were needed in quite large numbers, and machines for moulding these and other small articles developed slowly in the nineteenth century, together with machines for mixing and preparing sands and loam. Today the operation of a foundry without moulding machines would be difficult and for some foundries, such as those making motor-car parts, impossible. Some of the moulding machines now used are very complicated, and automatic or virtually so in operation, and the skilled, all-round, hand moulder is becoming a rarity.

For some purposes the process of casting itself has been mechanized. Cast iron pipes for water and gas mains, for example, are now very largely made by the centrifugal casting process devised by S. de Lavaud, a Brazilian, in 1914. In this process molten iron is poured into a revolving hollow cast-iron mould, which is kept in rotation until the casting is solid enough to be withdrawn. Centrifugal force throws the molten metal against the wall of the mould, where it solidifies and contracts in a few seconds. No core is needed, the mould can be used repeatedly, and the process produces a casting which is homogeneous and of better quality than if it had been cast in sand or loam. The first foundry for working the centrifugal casting process was put down at Stanton Ironworks, near Nottingham, in 1922. Centrifugal casting is now applied to numerous other types of work as well; cylinder liners and piston rings for internal combustion engines are typical examples.

Among the latest developments in the foundry are the CO_2

moulding process, shell moulding and the use of foam plastic patterns. The CO_2 process has developed extensively at the expense of dry sand moulding, and is even challenging green sand. Fundamentally the CO_2 process uses a special sand which is coated with some chemical such as sodium silicate. This is moulded in the normal way and CO_2 gas is then passed through it causing it to harden. There are variations of the process, but the basis is as stated.

Shell moulding, of German origin, has developed since the 1939–45 war as a means of making accurate moulds and cores quickly by machine. It also uses a special sand, but in this case it is mixed with a heat-hardening form of plastics powder. A relatively thin layer of this special sand is deposited on a heated metal surface shaped like the required mould, and there it hardens almost immediately. Quite complicated moulds can be built up from shells.

Even the realm of the patternmaker has been invaded by new materials. Traditionally the patternmaker has always worked in wood, though in the last fifty years or so metal has been used for patterns for high-repetition work. One of the limitations of the casting process is that since it is most suited to repetition work it can become expensive if only one casting is needed, as it may well be for, say, a special machine tool. Of late the practice has grown of substituting welded steel fabrications for large one-off or very short run castings, partly because of the cost of the pattern. In the last year or two the practice has developed of using stiff foamed plastic material for patterns for large one-off castings. This material is relatively cheap, it can be cut to the shape required and pieces can be stuck on if necessary. The mould is made round it as usual, but the pattern is not removed. When the molten iron or steel is poured into the mould, the foamed plastic pattern simply vapourizes.

Finally, mention must be made of a current re-use of an extremely ancient moulding method, the lost-wax process. This was known to the ancient Chinese, and it was employed on a small scale right down to the end of the nineteenth century for moulding statuary and other ornamental work of complex shape in bronze. Then it became virtually forgotten, but has been revived in a modified form in recent years to make small, complex castings for such things as jet engine parts. In modern lost-wax moulding a split metal mould is first made of the size and shape of the object required. This is then filled with molten wax, which soon solidifies, and the metal

28 Making a heavy shackle by hand forging

29 Forging a crankshaft under the steam hammer

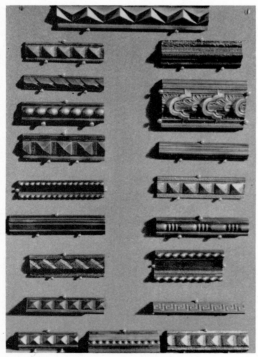

30 (*above*) A modern 3,000 ton hydraulic forging press, with preset control of stroke, at Walter Somers Ltd., Haywood Forge, Halesowen, Birmingham

31 (*left*) Some of the many shapes in which iron could be rolled, in addition to the ordinary bars and sections. These samples of fancy iron are in the author's collection

mould is opened for the wax model to be taken out. Sometimes, when the shape of the casting required is complicated, the wax pattern is built up of several pieces joined by heating and sticking together. Frequently several patterns are built up around a central runner like the branches of a tree. The next step is for the wax model to be coated with a refractory material (or 'invested'—hence the process is often called investment casting). Finally, the wax is melted out, leaving a hollow mould, which is then cast in the normal way. The process gives castings which are extremely accurate dimensionally and can be of any shape, regardless of whether or not the pattern could be taken out of an ordinary mould, for in this case the 'pattern', which is of wax, is melted out. Lost wax or investment casting is fairly slow and expensive, and is only used for small articles, but it is invaluable for such things as gas turbine parts, which are often of a difficult shape. For some gas turbine parts it is essential, for they are made of alloys which cannot be machined and must be cast exactly to finished shape and dimensions.

CHAPTER TEN

Smiths and Ironworkers

The art of the smith is an ancient and honourable one, and if the village blacksmith who did any kind of ironworking from making ornamental gates to shoeing a horse is now a rarity, the art itself is still both needed and widely practised, albeit with the aid of much machinery. In considering the work of the smith the main problem is to know where to draw the line, for the shaping by forging of iron (and later steel) has gone on and still goes on in many places wholly remote from the works where the metal is made. Moreover, there is now a widely dispersed industry making forgings by mechanical means (such as drop-stampings) in which the manual art of the smith plays no part. And there have been (and in some cases still are) specialized smiths making a single product such as nails or chains, sometimes using a small amount of simple machinery. It would be impossible in a book such as the present one to consider, even briefly, all the past and present variations of the smith's art. The definition chosen for the smith here is that he is the man who takes the iron or steel as made by the metal manufacturer, and by hammering or squeezing it while it is at a suitable heat, shapes it into some form useful as the raw material of some other trade. Such forgings were often called 'uses' in the nineteenth century and the term is still found here and there today. It covered such things as engine crankshafts, parts for ships, parts of railway locomotives; anything in fact which would be taken elsewhere for machining and otherwise finishing. It did not cover such things as nails, chains, chain cables, anchors or that wide variety of articles known by the expressive term 'oddwork'. These were all articles of use in their own right, ready to be used without further work.

Such, then is a forging for the purposes of the present chapter, though it is admittedly an arbitrary definition, and some of the items mentioned do in fact overlap into other classifications. Some things, too, started as pure smith's work and developed into an independent trade. Iron tubes are a case in point. They were originally made in very small quantities by blacksmiths, but special techniques soon had to be developed for making them in quantity. When Cornelius Whitehouse of Wednesbury, Staffordshire, in 1825 first drew a white-hot iron strip through specially-shaped dies, so turning it up into a circular form and welding together the meeting edges, all in one movement, he took tubemaking out of the hands of the smith and started the tubemaking industry. But our definition is sufficiently wide to enable us to look at the principles of forging and these are the basis, after all, of any forging process, whether simply manual or mechanized, general-purpose or entirely specialized.

A major characteristic of wrought iron is that when it is heated to a red heat (about 1,148°C) it becomes very soft and plastic, and can be bent, formed and shaped quite easily by hammering, squeezing or rolling (the latter is really only a form of squeezing). If two pieces are heated further, until they reach a temperature of about 1,300°C, they can then be welded together by the same process of hammering or squeezing. It does not matter whether the pieces are hammered or squeezed. So long as they are firmly pushed together at the correct temperature they will unite to form a homogeneous whole. All this was known at quite an early date, and both forging and welding were practised widely.

For a very long time hammering was the only process used. It is simple and for small forgings only a hand hammer is needed. For larger forgings the power hammer is desirable, in order to get the weld completed while the iron is still at welding heat. For if any attempt is made to weld iron at a lower temperature it will fail; the pieces of iron will not unite. But if a power hammer was a desirable tool it was still not essential, for it is quite feasible to complete a weld in two or more heats, welding, say half the joint, reheating the unwelded portion and then welding that. Sometimes it might even be desirable to do it this way, for the use of a power hammer, especially a tilt hammer, with uncontrollable force of blow, might well cause the finished product to be misshapen.

The welding of chain links will serve to illustrate the process; chains were certainly made by specialized smiths, but the welding was of a kind which could be done by any good blacksmith; indeed the blacksmith often had to alter a chain by cutting a link or several links out (which was called, simply, cutting) or to join one chain length to another by putting in a link (calling shutting, as many other types of weld were also called).

A small-diameter chain would be made by a blacksmith without any help. He would cut a piece of iron rod to the required length, heat it, bend it over his anvil to a U-shape, and then flatten out or scarf the ends of the U. Next, the blacksmith would thread the U through the previous link, and turn the ends of the U round until the two halves of the scarf overlapped. Finally, the scarfed part was put back in the fire and heated to welding heat, and withdrawn to be welded with a few quick blows of the hammer. The purpose of the scarf was to give a bigger surface for welding and to form it in a position where it could easily be hammered. If the chain were larger, the smith had the assistance of a helper or striker, and for the largest sizes of all, there were two strikers. In these larger sizes the weld was also done in two parts, the link being returned to the fire for reheating after the first half-weld was made.

Judgement of heat was of necessity a matter of experience, though it was not really as difficult as it sounds to tell welding heat from forging heat simply by looking at the iron. Where the smith's skill really did show up was in knowing just how long to leave the iron in the fire to get it to the right heat, for when it was pushed into the coke breeze the hot end was invisible. A learner always gave himself away by pulling the iron out two or three times to look at the colour of the hot end, which would be red at forging heat and almost white, beginning to sparkle, at welding heat. When the iron was at welding heat it had to be taken out of the fire and the weld made very quickly. To leave it in even a few seconds too long would be to burn it, while if the weld was made at below the correct heat it would be a failure. There was little margin.

It is only in comparatively recent years that there have been instruments for indicating and recording temperatures, and while they were and are used as an aid to steel forging (for which, in fact, they are now indispensable), they came too late to have any effect on wrought iron forging. So as long as wrought iron was forged it was done by the skill and judgement of the smith alone, and he had nothing in the way of machinery to help him but the power hammer. Nevertheless, some particularly fine work was done by smiths over a very long period, as is evidenced by surviving examples of both purely decorative work and of machine parts. Of decorative work museums can offer a good selection (though not all of it by any means is British), and there are particularly fine specimens of the work of Jean Tijou, at Hampton Court and St Paul's Cathedral. Tijou was a Frenchman, but spent twenty years in Britain in the latter part of the seventeenth century; and it was here that his best work was done.

Purely utilitarian wrought iron work can be seen in museums, too, though in this case the names of the smiths are not usually recorded. Many of the parts of the steam engines in the London and Birmingham science museums, for example, are the work of eighteenth and nineteenth century smiths. They are easily identified among the engine parts, for they are generally the long slender components, often bright polished, and very different in appearance from those of cast iron, most of which are relatively large and massive. The wrought iron parts are worth looking at both from the point of view of their fitness for purpose and because they often show that the smith could not resist the temptation to put a little decorative work on them. Much forged wrought iron, in fact, shows that joy in craftsmanship with which a good worker was so often imbued.

Because of its ease of welding quite large forgings could be made of wrought iron, although the iron itself was only available to the smith in the form of puddled balls weighing no more than about a hundredweight. Thus, such things as steam engine crankshafts and ships' anchors weighing several tons were built up from the required number of puddled balls. It is also characteristic of wrought iron that it is not very sensitive to the effects of heating and cooling, within wide limits, and in the making of complicated forgings it was possible to add small pieces here and there as required, by heating the larger mass only locally in a coke fire.

For this purpose the smith was usually provided with a floor area on which his forging could be placed and the fire built round the appropriate part. The small piece to be welded on was heated to welding heat in a hearth of normal pattern and carried in tongs quickly to the larger mass when it was ready. Meanwhile, the fire had been swept away from the larger piece, which had been turned to a more convenient angle if necessary. The small piece was then placed in contact with the larger one and welded there by hammering.

Joining two pieces of wrought iron by welding was far from being all that a good smith could do. He could bend the iron when it was hot, so making a crank, and he could twist one crank to a different radial position from the next. He could reduce the diameter of a bar in one or more parts of its length (or draw it down) or he could increase its diameter locally (or upset it). The purpose of drawing down is well exemplified by the production of a reduced end or tang on a tool.

Upsetting, on the other hand, was used where the major part of the forging was relatively small and only a minor part was relatively large. It would be far too slow and costly to take a piece of iron of the size of the larger portion and draw the bigger part of it down. So the smith took a piece of the size of the smaller portion and upset the end while it was hot. This could be done in more than one way. If the forging was a light one which could be held by hand, and if the upset was not too large in proportion, the easiest way was to heat the end and then, holding the iron in tongs, to bang the hot end hard on the anvil several times. When bigger forgings were being made, or when the upset was large in proportion to the rest of the forging, the upsetting was done by hammering by one or more strikers, or by a power hammer.

A very typical product made by upset forging is the ordinary screwed bolt as used by engineers, the head being upset from the shank. Such bolts are still upset, though today by a machine which shapes the head as a hexagon or square as it forges it. Smaller bolts are usually made by cold forging as well, but the principle remains unchanged.

Upsetting did not have to be at the end of the forging. The whole principle depended on the fact that the hot part was soft and plastic and the remainder was stiff and unyielding. So if a small part of the centre of a bar was heated and the bar was then bumped heavily on the anvil, the heated part would swell out into a spherical shape. A typical use for this variant of upsetting, known as jumping, was the stanchion for a handrail. This had an upset ball at the top and a jumped ball in the middle.

A good smith could also carry out an operation which can best be termed 'carving' hot iron though that is not the technical term for it; there is no specific technical term for the cutting of hot iron with a chisel or gouge, just as a carpenter would cut a piece of wood. But the smith would use a chisel or gouge, held in a handle made of iron rod, while his striker gave it the necessary blows with a sledge hammer. In this way he would cut recesses such as were required in the head of a heavy chain swivel, and it was the boast of a good smith that this work needed no machining when he had finished it.

Sometimes it would be necessary to turn quite a thick or large-diameter piece of iron through 90° or even 180° round a small radius, and this posed a problem, for the outer part of the bend would stretch and cause the iron to thin out locally. It would not tear; the hot iron would stand the stretching but the finished component would be undersized at that point. This could happen in many forgings, a typical example being a large swivel used for ships' mooring chains or cables. (A cable, it should be explained, is a chain with a central stud of cast iron in every link. The studs prevent the links from twisting and cause the cable to run out freely when it is released. It was introduced by a British naval officer, Captain Samuel Brown in 1808.)

To overcome the trouble a V-shaped cut was deliberately made in the outer side of the iron before it was bent. This cut opened out as the bend was made. It was closed by welding-in a piece of iron of V-shape, and the forging was then of the correct size and shape everywhere. This V-shaped piece was called a glut. Very often the smith made his own gluts from any piece of iron that happened to be handy, but it was also possible to get V-shaped bar or glut iron from the rolling mills in various sizes. This had only to be cut off to the required length.

So a smith could forge quite complicated shapes with very little equipment and nothing of a mechanical nature except a power hammer. Even this was dispensed with sometimes, the smith either having a fairly large number of strikers or making use of a device known as the 'beetle'. This was very old indeed, being known to the ancient Chinese. It was simply a heavy weight which was hauled up on a rope or chain over a pulley in the roof of the forge, and allowed to fall on the forging. Four men with ropes attached to the beetle in cruciform plan directed its fall to the point required. In its earliest versions the beetle was lifted by manpower, but later, when steam power became available, a steam winch was used. Beetles, however, were not common, and until the coming of the steam hammer the smith's only tool other than the hand hammer was generally the tilt or helve hammer.

The distinction between the nose or frontal helve and the belly helve has already been noted. It was the belly helve which found most favour in the heavy forgings trade, for it had an unrestricted area round the hammer and anvil which was very useful when a forging had to be manoeuvred to get the hammer blows in the right place. This manoeuvring often had to be done quickly and the only equipment the smith had to help him was at best a swivelling jib crane with a manually operated winch on it. Sometimes he did not even have that; the whole forging was pulled and pushed and turned over manually. To help the men heavy levers would be bolted on to the forging temporarily, and when the shape of the forging was suitable the outer end of it, away from the hammer, was wedged into a heavy component (itself a forging) called a porter bar, which helped to balance the forging and served as a means of applying leverage to it to move it.

The first development in forge machinery came when Nasmyth invented the steam hammer in 1839. This has already been mentioned as a means of hammering or shingling puddled balls, but it is of particular interest in the present context because it was developed for making wrought iron forgings, not for use in the ironworks; it was adopted for that purpose later. The steam hammer was Nasmyth's answer to the problem of forging a wrought iron shaft of 30 inches diameter for the steamer *Great Britain*. Nasmyth was already engaged in supplying machine tools for making the engines of this great (for its day) steamship, and the engineer in charge told him that no forge firm would quote for the shaft as none had a hammer big enough. Nasmyth's idea was to build a framework over an anvil to carry a steam cylinder from the piston rod of which a heavy hammer or tup would be suspended. Steam would raise the piston and so lift the hammer, and, when the steam was cut off and the cylinder opened to exhaust, the hammer would fall by gravity. It was a simple and effective idea, which later became a great success, but no hammer was made for the *Great Britain*'s shaft, for a change in plans gave the ship different engines and the 30-in shaft was not needed.

The hammer remained for the time being just an idea in the inventor's sketch book, where it was seen by a visiting Frenchman, and translated into the first actual hammer in the ironworks at Le Creuzot, France. However, Nasmyth obtained a patent for the hammer in 1842, and soon afterwards he made a considerable improvement in the design, which was to render the steam hammer of great value to the smith.

This was the double-acting design, which was introduced in 1843. Steam was now admitted above the piston to drive it down and so give a greater force than gravity alone could do. With the first single-acting hammer excellent control of blows was possible, for by keeping steam under the piston the hammer would stay up, and the blow could be delayed as long as necessary; thus the position of the forging could be altered as required between blows. By letting the steam out gradually through a manually controlled valve, the hammer could be made to fall as slowly, and therefore as gently, as the smith wished.

With the double-acting hammer the same, or an even better, control of the descent of the tup was possible, in addition to a blow of increased force. A simple arrangement of valves enabled the steam to be 'balanced' on the two sides of the piston, so holding the hammer stationary. By varying the balance of pressure the hammer could be made to lift or fall at any desired speed. For the maximum blow the piston was raised to the top of the cylinder, the steam space below the piston was opened to atmosphere, and at the same time full steam pressure was admitted to the top of the piston.

So delicate was the control on a steam hammer that it was possible to place an egg on the anvil and bring the hammer (which may have weighed anything from a hundredweight or so to several tons, according to the purpose for which it was intended) down on to the egg so gently that it was cracked, but not shattered. Indeed, this trick was sometimes performed to impress visitors. It was also possible to modify the valve gear in such a way that the hammer, once set, would stroke continuously until stopped manually. This had certain limited uses, as in the drawing down of bars, and the hammer driver would then merely hold the control open while the hammer worked automatically, until the operation was finished. For the most part, however, the intricacies of forging called for a great variety of blows, and the control was usually wholly manual.

So the steam hammer increased the versatility of the smith, not only in the matter of forging size, but also in the complexity of forgings and the speed with which they could be made. For although the steam hammer was originally designed to deal with the problem of making a forging which was too large for the hammers of the time, it was the ease of control which was the steam hammer's greatest virtue. It became in effect an extension of the hand hammer which had always served (and still serves) the smith so well, and increased the range of work he could do.

Mild steel made its presence felt in the traditional iron making and rolling trades before it had any effect in the forges, but it arrived there eventually, as it was bound to do. It had its advantages from the smith's point of view, for it could be obtained in fairly large pieces and the rather laborious business of building up a large piece of iron for a heavy forging from puddled balls was eliminated. It had its disadvantages, too, especially as time went on and carbon and alloy steels came to be forged, for these steels are far more difficult to handle. Temperature control is much more important; some steels do not permit of liberties. They must be dealt with at the correct temperature and the forging must often be done in a particular way. Moreover, as the size of steel forgings became larger, so steam hammers, even the biggest, were incapable of applying enough force.

Steel for forgings, like steel for other purposes, met with some opposition at first. But the development of steel, as opposed to iron, shipbuilding, and the increase in the size of ships, led to a demand for larger forgings, and by the 1870s the steam hammer was inadequate for some types of work. So the heavy hydraulic press began to come into its own.

Not only is there the possibility with a hydraulic press of exerting a much greater force than can be obtained from the steam hammer; there is a fundamental difference in the forging effect of the two machines. The energy of a steam hammer blow is largely expended superficially, which is unimportant when the forging is relatively small, but when the forging is large the force of the blow would be expended before it reached the centre. Consequently the centre would not be properly worked. With the hydraulic press, on the other hand, the force is applied relatively slowly and steadily, and it penetrates right to the centre of the metal being forged.

The characteristics of a hammer blow were well understood by Bessemer, and he had ideas for a type of forging press, but nothing effective was done in his time. Development of the press for forging was to come at a later date. Neither the history nor the principles of the hydraulic press need be discussed here; the subject is one for a history of engineering. It is sufficient to say that hydraulic power was known and used late in the eighteenth century, and was applied for various purposes by the middle of the nineteenth. W. G. Armstrong made considerable improvements to hydraulic machinery in 1851, and in 1865 Joseph Whitworth used a hydraulic press for consolidating molten steel and for forging the resulting ingot. A hydraulic press with a force of 2,000 tons was installed in Sheffield in 1863 for working armour plates, and a number of others of similar size followed.

British machinery manufacturers played a considerable part in the development of the hydraulic press for forging steel, and by the end of the nineteenth century had achieved an enviable reputation for their presses. In 1905 an American firm took a licence to manufacture hydraulic forging presses from Davy Brothers of Sheffield, and development of the press has since taken place on both sides of the Atlantic, and in Germany and other countries as well.

All the early hydraulic forging presses (and for that matter other hydraulic machinery) used water as the hydraulic fluid. The press had a large cylinder into which water was forced under pressure, causing the press ram to descend on the forging. There were variations in the design. Some presses had more than one cylinder, and some took their pressure from a hydraulic accumulator (the invention of Armstrong) which was itself charged by steam-driven hydraulic pumps. Others got their pressure from a direct steam/hydraulic apparatus, in which a large steam piston working at relatively low pressure was connected directly to a smaller hydraulic piston which generated the necessary high hydraulic pressure. But the effect in all cases was the same. Very great force, amounting to several thousand tons, could be brought to bear on the heated forging. Control was by a hand lever operating hydraulic valves.

Handling gear developed with the presses, which were now capable of making forgings far too large to be manhandled and beyond the capacity of simple jib cranes. So the overhead travelling crane was brought into the forge, and with the advent of electric drives an electric turning gear was devised. This consisted of a very heavy chain which depended in the form of a loop from the turning gear; this, in turn, was suspended from the crane hook. The forging was suspended by its porter bar or by its own extended tail in the chain, and the crane driver, by using his electrical controls, could cause the chain to rotate the forging. He could also, of course, raise and lower the crane hook and with it the forging, and traverse the crane hook along or across the shop as required.

This sort of equipment, with presses ranging in power from a few hundred to about 8,000 tons, was well known and fairly widely spread by the early years of the twentieth century. It would not be true to say that it was common, for the production of heavy steel forgings was always on a relatively small scale compared with steel-making. But wherever heavy forgings were made such equipment was to be found. It lasted, with little change, until recently, and served to produce some very fine forgings. Ships' propeller shafts, boiler drums for super power stations, and shafts for electric turbo-alternators are typical examples.

From about the early years of the present century until the end of the 1939–45 war there was little to report in the design or operation of steel forges. Some improvements in plant detail were made, and all the time, as alloy steels came into greater prominence, methods of heating (and of heat treatment after forging) called for better equipment and more precise control. A gradual change in forge furnaces had been taking place for many years. In the old days of wrought iron, coal-fired furnaces were adequate, but steel demanded better control of heating than a coal furnace, manually fired, could give. A change to producer-gas firing, using the same sort of gas producers as on open-hearth furnaces was made in some forges early in the present century, and this gave at least a measure of accurate heating control. The use of instruments to indicate (and often to record automatically) temperatures in furnaces and various other operational variables as well, grew with the need for better process control, but the forging presses themselves showed little sign of change. Massively constructed and relatively slow moving, as they were, it was virtually impossible to wear them out, and many of them worked for half a century and more. But if they did not wear out, they certainly became out of date, for research was carried out in the forging branch of the iron and steel industry, as well as in the other branches, and as new mechanical and electrical devices became available, so the forging machinery makers put them to use. Besides, in the forges, as elsewhere, all labour was becoming expensive, and highly skilled labour scarce.

The 1939–45 war, with its demand for forgings of great precision, sometimes in unfamiliar alloys, provided an incentive to develop new plant. Of course nothing much could be done while the war was on, but when it ended, some quite novel apparatus began to appear. A leading organization in the development of forging methods and machines in the post-war years has been The British Iron and Steel Research Association, and it has been well supported by plant manufacturers.

Three major changes have taken place in the forges as a result of development work; the introduction of oil-hydraulic presses, automatic stroke-length (or thickness) control, and integrated forge and manipulator operation. From the earliest days until recently the fluid used in hydraulic presses was water. But the development of high-speed electrically driven pumps, together with suitable mechanical and electromechanical control valves made possible the use of oil as the hydraulic medium, which makes the press self-contained, versatile, and high-speed if needed.

The first oil-hydraulic forging press in Great Britain was installed by Walter Somers Ltd, Halesowen, near Birmingham, in 1958. It was relatively small as forging presses go, having a capacity of 500 tons, but it was of advanced design, and could work at 130 strokes/minute when required. It introduced, too, the new conception of thickness control. It should be explained that one of the difficulties in press (or for that matter, hammer) forging under manual control is to avoid over-forging, or making the article being forged thinner than it ought to be. So the press operator, when getting near to the required dimension, would forge slowly and rather gently, to make sure he did not go too far. This wasted time and, probably, heat as well.

With the thickness control on the Somers press it is possible to set on a dial the required distance between the forging press top tool and the anvil or bottom tool, and the press will stop automatically at this point, within $\frac{1}{64}$ in. This has only been made possible by the development of extremely fast-acting hydraulic valves and electric and electronic controls which take into account not only the movement of the press ram, but also the amount by which the press columns stretch under load. Thickness control has proved extremely successful, and all modern forging presses incorporate it. A much larger press, of 3,000 tons capacity, with thickness control, was installed at the same works in 1961. This, too, proved so successful and versatile that it replaced not only an older press of the same capacity, but two others as well, doing the work of all three.

Presses of such speed and accuracy call for something more than an overhead crane to carry and manipulate the forging, and the mechanical manipulator, itself quite a complicated machine, is an essential auxiliary. A simple form of manipulator, steam driven, had appeared at the Vickers Works, Sheffield, as early as 1897, and soon after 1900 others were in use in America (where they are often said to have originated). But the first in what might be called the modern form was made and used by John Baker at Kilnhurst Steelworks, near Rotherham, in 1925. It consisted of a travelling carriage mounted on rails, and carrying a massive pair of jaws on a horizontal shaft which could be elevated and rotated by a combination of hydraulic and penumatic power. A forging of up to 30 cwt in weight could be held in the jaws, lifted, turned, and moved backwards and forwards as it was forged, all by mechanical power.

This remains the basis of the forging manipulator of today, though the equipment, all-hydraulic now, is much more complicated and faster in operation, and can be made to carry forgings weighing many tons. Forgings are now made as a matter of routine by two operators, one controlling the press and the other the manipulator.

The next obvious step is to integrate the operation of press and manipulator, and control both from a single panel. This has been done, the first installation being put to work at Parks Forge Ltd, Wigan, in 1964. In this forge an 800-ton press and a rail-mounted manipulator are fully integrated, and the movements of both can be set on a series of dials on a control panel. When this is done, the operator has only to move a single lever, and the press and manipulator will go through the preset sequence automatically.

Nor is this all. Development work done by The British Iron and Steel Research Association has proved the feasibility of programming a press and manipulator in such a way that they can be controlled automatically by punched cards or tape. It is then only necessary to translate the drawing of a forging into a card or tape programme, feed this into a reading machine, put a heated ingot into the manipulator jaws, and press a button. The machine will do the rest, and stop when it has done it. No such plant is yet in commercial operation, but it seems more than likely that one soon will be.

Forging, like all the other branches of iron and steel manufacture and manipulation, has gone a long way since primitive man first forged by hand his little pieces of iron.

Gazetteer

So much of interest to the technical historian has been carelessly or even wantonly destroyed over the years (some in the lifetime of the present writer) that there is little enough to be seen generally, and some periods and processes are entirely unrepresented. Moreover, some of what has been preserved is not yet open to the public and the fate of other relics is in the balance. A few relics remain at present which are so incomplete or badly damaged that there is no point in trying to preserve them. Hinkshay and Old Park blast furnaces in Shropshire are in the latter category. They are on the site now being developed for Dawley new town, and will in due course disappear. All that can reasonably be done is to make measured drawings of what remains and this is in hand.

Fortunately, however, some relics do remain and a few, at least, are very fine indeed. They are listed below under subjects. It should be noted that some of these relics are on private property and that permission to visit them should be obtained from the landowners or their agents. This is usually granted to responsible persons; it is against the irresponsible that the owners are naturally on their guard and, regrettably, experience has taught some owners to regard all visitors as in the irresponsible class until proved otherwise.

BLAST FURNACES

The best example of an eighteenth-century blast furnace site is that at Coalbrookdale, Shropshire, which is preserved, together with an indoor museum, by Allied Ironfounders Ltd, and is open to the public at stated times. It has two blast furnaces in a fair state of preservation, though incomplete in the hearth and throat and without blowing apparatus. There is sufficient on the site to enable the visitor with a reasonable knowledge of blast furnace operation to reconstruct, mentally, the method of working the furnaces. The indoor museum has an excellent collection of castings, illustrations, and other items showing the work of Coalbrookdale from the time of Abraham Darby I to the present day.

Other good examples of seventeenth and eighteenth century blast furnaces can be seen at Charlcot, Neenton, Shropshire; Bonawe, near Oban, Argyllshire, Scotland; Argyll, Loch Fyne, also in Argyllshire; Duddon, near Millom, Cumberland; Morley Park, near Heague, Derbyshire; Rockley, near Barnsley, Yorkshire; Cockshutts, also near Barnsley; Whitecliff, Forest of Dean, Gloucestershire; and Eglwysfach, near Dovey Junction, Cardiganshire.

There is now no typical nineteenth-century blast furnace in existence.

FORGES

Pride of place goes to Abbeydale, Sheffield. This is a complete industrial hamlet of the eighteenth century. It contains six crucible steel melting furnaces, two tilt hammers, grinding shop, hand forges, warehouses and workmen's cottages, and was used for the manufacture of scythes until the 1920s. All the machinery was normally water driven, though a small steam engine was provided as a reserve source of power. After lying derelict for many years the works became the property of Sheffield Corporation. It is now in the process of being restored to full working order by the Council for the Conservation of Sheffield Antiquities. When complete it will be opened to the public as a museum in the care of Sheffield Corporation.

Wortley Top Forge, near Stockbridge, Sheffield, the remaining part of Wortley Ironworks, is preserved by the Sheffield Trades Historical Society. It is on an ancient site, but the buildings and machinery there are eighteenth century. The equipment includes water wheels, two tilt hammers and a very interesting little late eighteenth-century two-high rolling mill formerly at Wortley Low Forge, nearby. Other items of ironworking interest from the district have been brought to Wortley for preservation.

A private museum in the Kirkstall, Leeds, works of Kirkstall Forge Engineering Ltd, also contains two tilt hammers.

WROUGHT IRON WORKS

No puddling or mill furnace has been preserved. There is a small helve hammer at Abbeydale, but it is not a typical ironworks forge hammer. No complete rolling mills have survived, but there is part of a nineteenth century type guide mill in store at Staffordshire County Museum, Shugborough, near Stafford. This is to be on display in the public section of the museum at, it is hoped, an early date. The National Museum of Wales, Cardiff, has some sheet rolling mill exhibits.

MISCELLANEOUS

Birmingham Museum of Science and Industry has a good example of a tilt hammer formerly used in a local edge-tool works. There are models of interest at several museums including Cardiff, Stafford and the Science Museum, London. In the London collection there are some good models showing more modern plant which would be far too large to display. Bilston Art Gallery and Museum, Bilston, Staffordshire, acquired in 1966 the main historical exhibits from the Staffordshire Iron and Steel Institute centenary exhibition held in that year, and has them on show to the public. This collection, presented by private individuals and local firms, is of models, drawings, photographs and specimens relating to the Black Country iron industry in the seventeenth, eighteenth and nineteenth centuries. Museums, including small local ones, are always worth visiting, for they sometimes have a specimen or two which is of interest, though they may not possess an iron and steel collection as such. They often contain drawings and photographs of old iron and steel plant which are little known elsewhere.

A Select Bibliography

THE EARLY PERIOD

Wealden Iron, Ernest Straker (G. Bell, 1931), is the classic guide to the iron trade of south-east England from its earliest days to its final extinction. It is not deeply technical, but is thorough, has a valuable list of furnace and forge sites, and deals with products as well as processes.

History of the British Iron and Steel Industry, H. R. Schubert (Routledge & Kegal Paul, 1957), covers the period from *c.* 450 B.C. to A.D. 1775. It is a work of considerable scholarship, and quite detailed on technical matters.

COKE SMELTING AND WROUGHT IRON

Dynasty of Iron Founders, A. Raistrick (Longmans, 1953), is the standard work on Abraham Darby and his successors at the Coalbrookdale Works. The subject is treated both technically and economically, and Darby family matters are also given full consideration.

Quakers in Science and Industry, A. Raistrick (Bannisdale Press, 1950), covers the period between 1650 and 1800, and although it deals, as its title indicates, with Quaker industrialists and scientists generally, has useful chapters devoted to ironmasters; there were several Quakers in this industry besides the Darbys.

Iron and Steel in the Industrial Revolution, T. S. Ashton (University Press, Manchester, 1924, reprinted, 1963) is the classic economic study of the subject. It surveys the charcoal iron industry briefly, and deals in greater detail with coke smelting, wrought iron and the effects of mechanical power. The study ceases with the beginning of the nineteenth century.

SPECIFIC AREAS

There are numerous local industrial histories which deal with iron and steel to a greater or lesser degree. These include such works as *Furness and the Industrial Revolution*, J. D. Marshall (Barrow-in-Furness Library and Museum Committee, 1958); *The Industrial Revolution in North Wales*, A. H. Dodd (University of Wales Press, Cardiff, 1951); *The Industrial Development of South Wales*, A. H. John (University of Wales Press, Cardiff, 1950); *Men of Iron*, M. W. Flinn (University Press, Edinburgh, 1962); and many others. But they are almost all concerned principally if not entirely with economic and sociological history. An exception is *The Black Country Iron Industry: A Technical History*, W. K. V. Gale (The Iron and Steel Institute, 1966), which is devoted to technical developments.

NINETEENTH CENTURY

There are one or two classics of particular value to the technical historian because they were written to detail the latest technical practice of the time. These include *Iron, its History, Properties and Processes of Manufacture*, W. Fairbairn (A. & C. Black, 1861) and *Metallurgy: Iron and Steel*, J. Percy (John Murray, 1864). The former is not too technical but the latter, which is the great classic of its period, is a massive volume of over 900 pages devoted almost exclusively to technical matters. It is of great value to the technical historian but is not an easy book to read, especially as some of the terms used have now been superseded. A later book of the period *The Metallurgy of Iron*, T. Turner (C. Griffin, 1900), though technical, is better for the general reader.

Guide to the Iron Trade of Great Britain, S. Griffiths (Privately published by Griffiths, 1873), is a well-known volume of general interest, but it is inaccurate historically and is badly arranged. It contains some good engravings of ironworks, and the trade advertisements, which are numerous, are informative.

SPECIFIC ASPECTS OF IRON AND STEEL HISTORY

Literature on particular sections of iron and steel history is rather sparse, but there are some useful works including *Autobiography*, H. Bessemer (Engineering, 1905); *The Story of the Mushets*, Fred. M. Osborn (Thomas Nelson, 1952); *Life of Sir William*

Siemens, W. Pole (John Murray, 1888); and *History of the Iron, Steel and Tinplate Trades of Wales*, C. Wilkins (Williams, 1903). *Sidney Gilchrist Thomas*, L. G. Thompson (Faber & Faber, 1940), deals more with the personal life of Thomas than with his work but, written by his sister, is authoritative and informative. *Knotted String*, Harry Brearley (Longmans, 1941), is disappointingly uninformative on stainless steel, the major work of its author. His *Steel-Makers* (Longmans, 1933), though again uninformative on stainless steel, has a fine description of crucible steelmaking.

MODERN HISTORY

History of the British Steel Industry, J. C. Carr, W. Taplin, and A. E. G. Wright (Basil Blackwell, 1962) is an exhaustive work which, after a brief introduction to the pre-Bessemer period, studies the history of iron and steel from Bessemer to 1939. An epilogue takes the story down to 1960. The book deals equally with economic and technical matters. *The Development of the Modern British Steel Industry*, B. S. Keeling and A. E. G. Wright (Longmans, 1964) can be considered as complementary to the volume mentioned above. It deals with the period 1920 to 1964, largely on an economic basis, but technical matters are not overlooked.

COMPANY HISTORIES

Company histories range from the privately published brochure to the commercially published volume, the latter being few in number while the former are often hard to find and frequently inaccurate. Published volumes include *Carron Company*, R. H. Campbell (Oliver & Boyd, 1961) and *Vickers, a History*, J. D. Scott (Weidenfeld & Nicolson, 1962); this deals with all the activities of the Vickers group, but has useful sections on steel.

MISCELLANEOUS

The *Transactions of the Newcomen Society*, the *Journal of The Iron and Steel Institute* and the *Bulletin of the Historical Metallurgy Group* of The Iron and Steel Institute all contain papers on various aspects of iron and steel history.

Index

Abbey Steelworks, 100
acidic slag, see slag
Adamson, Daniel, 73
Alleyne, Sir John, 79
   and Alleyne's traverser, 79, 80
   and Alleyne's reversing mill, 80
Allied Ironfounders Ltd, 26
all-sinter burdens, 99
Appleby-Frodingham Steelworks, 99
Armstrong, W. G., 133
Automatic gauge control (AGC), 105–6

Baker, John, 136
Barlow, H. W., 79
Barrows and Hall, 55
beams, 88, 89
Bedson, George, 81, 82
beetle, the, 129
Belgian mill, see rolling mills
bell and hopper device, 54
bellows, 13, 14
   water-driven, 19, 36
beneficiating, 98
Bessemer, Henry, 69, 71, 91, 99, 103, 133
   and Bessemer process, xv, 71, 73, 74, 77, 78, 79, 80, 95, 100
   and 'Bessemer iron', 73
   and Henry Bessemer and Company, 73
best iron, see irons
Blackband ores, see iron ores
Blaenavon Company, 72, 78
blast furnaces, 67, 75, 95, 97, 98, 100, 112, 118
   description of, 14–8
   coke fired, 26–9, 35
   increase in size of, 28, 35
   modern plant, 35–8
   and hot blast, 41–2
   and effects of hot blast, 43
   and introduction of circular hearth, 44–6
   and use of cinder in the charge, 46
   and use of waste gas, 53–4
   and the closed forepart, 83
   fast 'driving' of, 85, 87
   and pig-casting machine, 85, 95
   and tap-hole gun, 85
   mechanically charged, 85–7, 95
blast furnace metal, see cast iron
blister steel, see steel
Blochairn Steelworks, 88
bloomery, see furnaces
boil, the, 57, 58
Bolckow, Vaughan & Company, 79
bosh cinder, or bosh slag see slag
Box, Geoffrey, 21
Boyden, Seth, 119
box pile, see pile
Brearley, Harry, 90
Britannia Steelworks, 88
British Iron and Steel Research Association, 101, 105, 111, 135, 136
British Standards Institution, 89
broadsiding, 64
Brown, John, 73
Budd, J. P., 54
bulk steelmaking, see tonnage steelmaking

bulk steels, *see* tonnage steels
'bull-dog', *see* roasted cinder
Bulmer, Bevis, 21
Butterley Ironworks, 79

Cammell, Charles, 73
carbon steel, *see* steels
carbonate ores, *see* iron ores
case-hardening, 8
casting, 112–23
castings, xii, 9, 116, 118, 119, 120, 121, 122, 123
cast iron, 3, 6, 9, 12, 18, 26, 31, 32, 42, 43, 69, 112, 113, 114, 115, 119, 120, 121, 127
    malleable cast iron, 119–20
cementation process, 70
chafery, 20
charcoal,
    early use of, 3
    shortage of, 22–3
    furnaces, end of, 38
cinder, *see* slag
cinder pig, *see* part-mine iron
clay gun, *see* tap-hole gun
coal, difficulties of using as fuel, 23–5
Coalbrookdale foundry, 26, 28, 29, 30, 31, 32, 36
    Company, 38
coke smelting, development of, 27–9
cold rolling of iron, 63
'come to nature' process, *see* puddling
common iron, *see* irons
computers, 106, 107
concrete, xi
Consett Iron Company, 56
continuous casting, 102–5
continuous mill, *see* rolling mills
continuous wide strip mill, *see* rolling mills
continuous wide strip process, 92–4, 95
Cort, Henry, 32
    and reverberatory furnace, 33
    and dry puddling process, 34
    and grooved rolls, 35
Cowper's stove, 42, 75

Cranage, Thomas and George, 32
cropends, 56
cross pile, *see* pile
crown iron, *see* common iron
crucible steel, *see* steels
cupolas, 118–9

Darby I, Abraham, 26
    and Coalbrookdale foundry, 26, 38
    and coke smelting process, 26–9, 32
    and iron foundry practice, 115–6
Darby III, Abraham, 31
Davy Brothers of Sheffield, 133
de Lavaud, S., 121
degassing, methods of, 109–11
Diderot and D'Alembert, *Encyclopédie*, xvi
direct reduction method, 11–4, 97
double shear steel, *see* steels
Dowlais Ironworks, xvii, 72
dry puddling, *see* puddling
Dudley, Dud, 24–5, 116
    and *Metallum Martis*, xvi 24
Dufrenoyetal, *Voyage Méttallurgique en Angleterre*, xvi
Dwight, A. S., 87

Ebbw Vale Ironworks, 72, 73
Ebbw Vale Steelworks, 93, 95, 100
electric power, use of, 84, 87, 91, 92, 96, 97, 105, 133, 134
electrolytic tinning process, 94, 95
electronics, use of, 97, 105
electroplating, 95
electro-slag refining process, 109
Engineering Standards Committee, 89

'fancy' irons, *see* irons
fettling, 49, 56, 57, 58, 77
finery, 14
    description of, 18, 19
    inefficiencies of, 32
    disappearance of, 38
Fletcher Solly and Urwick, 83

INDEX

flushing, *see* slagging
Foley, Richard, 21
forge train, *see* rolling mills
forging, 112, 124–37
forging manipulator, 136–7
forgings, xii, 9, 59, 124, 125, 127, 128, 129, 130, 131, 132, 136
Fritz, John, 52
Frodingham Ironworks, 86
fuel economy, 50–1, 75
furnaces
    direct-reduction (bloomery), 12–4
    blast, *see* blast furnaces
    finery, 14, 18–9, 32, 38
    reverberatory, 33, 118
    refinery, 34
    mill, 50, 61
    puddling, 55–8
    regenerative, 75
    open-hearth, 76, 77, 91, 95, 99, 101, 102, 134
    electric, 91, 92, 96, 107, 108, 119
    annealing, 93
    vacuum, 108
    air, 118
    crucible, 120

gagging-up, 59
gauges, 63, 64
German Rotor process, 101
Gibbons, John
    and circular hearth, 44–6
    and part-mine iron, 46
Gilchrist, P. C., 78
Grey, Henry, 88
Griffiths, S. *Guide to the Iron Trade*, xvi
guide mill, *see* rolling mills

Hadfield, Sir Robert, 89
    and manganese steel, 89–90
    and silicon steel, 90
Hall, Joseph, 46, 50, 56
    and wet puddling process, 46–9
    and roasted cinder, 49
hammers,
    hand, 19, 125, 130, 132
    tilt, 19–20, 59, 125, 130
    helve, 19, 59, 60, 67, 130
    power, 21, 125, 126, 129, 130
    steam, 59, 60, 130–2
Hanbury, John, 35
heat, a, 56, 57, 58
helve hammer, *see* hammers
hematite irons, *see* irons
hematite ores, *see* iron ores
Hero of Alexandria, 68
Heroult, Paul, 91
hot saw, 67–8
hot-blast stove, 42, 75
Huntsman, Benjamin, 70, 89
    and crucible method, 70
hydraulic forging press, 132–3, 135, 136

indirect reduction method, 14–20
Iron in the Making, xvii
iron ores, 1–3, 37, 97, 98, 99, 119
    magnetite, 2
    hematite, 2, 3
    low grade, 2
    lean, 2
    carbonate, 2
    Blackband, 41
iron-bearing sands of New Zealand, 2
Ironbridge, 31
ironmaking materials, world's resources of, xi
irons
    pig iron, xii, 6, 18, 20, 33, 34, 38, 40, 46, 47, 48, 56, 58, 71, 72, 73, 76, 77, 78, 79, 85, 118
    cast iron, *see* cast iron
    meteoric, 1
    pure, 2
    white, 34
    part-mine, 46
    common, 61
    best, 61
    'fancy', 62
    hematite, 73
    phosphoric, 78

joists, xv, 63, 80, 89

INDEX

Junghans, S., 103

Kilnhurst Steelworks, 136

LD process, 100–1
lean ores, *see* iron ores
Legge, B., *Guide to the Iron Trade*, xvi
Lloyd, R. L., 87
LNWR, 73
Loewy-Robertson 'Constant-Gap' mill, *see* rolling mills
looping, *see* Belgian mill
low grade ores, *see* iron ores
Lucas, Samuel, 119
Lührmann, F., 83

magnetite ores, *see* iron ores
malleable cast iron, *see* cast iron
manganese steel, *see* steels
Martin, Emile and Pierre, 76
mechanical traversers, 79–80
merchant iron, *see* common iron
metallurgy, 74, 89, 106, 112
meteoric iron, *see* irons
mild steel, *see* steels
mill scale, 49, 56, 57
moulding machines, 121
moulding materials, 116–18
moulding processes, 116–18, 121, 122
muck bar, 60, 61, 68
Mushet, David, 40–1
 and *Papers on Iron and Steel*, xvi 40
 and Blackband ironstone, 41
Mushet, Robert Forester, 40, 73
 and tungsten alloy steel, 74, 89

Nasmyth, James, 59, 130
natural gas, 97
Naylor, Vickers and Company, 120
Neilson, James Beaumont, 41
Neilson, Walter, 41–2
Newcomen's engine, 30, 31

Oakes, T., 45, 46

oil-hydraulic forging press, 135
Onions, Peter, 32
open-hearth process, 74, 76–8, 79, 87
oxygen steelmaking processes, 99–102

Park Forges Ltd, Wigan, 136
Parry, George, 54
 and bell and hopper device, 54
part-mine iron, *see* irons
Percy, Dr John
 and *Metallurgy of Iron*, xvi 74
phosphoric irons, *see* irons
pig boiling, *see* wet puddling
pig irons, *see* irons
pig-casting machines, 85, 95
pile, a, 61
 plate pile, 61
 box pile, 61, 62
 cross pile, 62
plates, xii, 63, 64, 66, 88, 96, 103
power hammer, *see* hammers
puddled bar, *see* muck bar
puddling,
 and 'come to nature' process, 34, 58
 process of, 55–8
 dry puddling, 34, 38, 46–7, 48, 49
 wet puddling, 46–9
 furnaces, *see* furnaces
pure iron, *see* irons

rabbling, 57
Ramsbottom, John, 80
Rastrick, John Urpeth, 50
 and waste-heat boiler, 50–1, 75
raw coal smelting, 43
Réaumur, xvi 69, 119
refinery, 34
regenerators, 75
roasted cinder, 49
Rogers, S. B., 47
rolling mills, 55, 56, 60, 65–7, 87, 96, 102
 slitting, 21–2, 35

INDEX

rolling mills-*continued*
   two-high, 51–2, 60, 63, 66, 67, 80
   three-high, 51, 52, 63, 80
   guide, 51, 52–3, 67
   forge train, 60, 63, 67
   Alleyne reversing mill, 80
   Ramsbottom's reversing mill, 80
   Belgian mill, 80–1
   continuous, 81–2
   continuous wide strip, 92–4
   Loewy-Robertson 'Constant-Gap' mill, 105
   automatic operation of, 107
J. Rovenzon, *A Treatise of Metallica*, xvi

Savery's engine, 30
Scotch tuyere, 43
shear steel, *see* steels
sheets, xii, 63, 64, 65, 66, 92, 93, 96, 103
Shelton Iron and Steelworks, 104
shingling, 58, 59, 60
Siemens, C. W., 74–5
   and open-hearth process, 74, 76–8, 121
   and Siemens-Martin process, 76
   and 'Sample Steelworks', 76
   and Landore Siemens Steel Company, 76
   and arc furnace, 91
silicon steel, *see* steels
sintering, 98, 99
slag, 17, 70, 77
   presence in wrought iron, 5
   slag notch, 37, 38
   acidic slag, 48
   bosh slag, 47, 48, 49, 56
   as agricultural fertiliser, 79
slagging, 37–38
slitting mill, *see* rolling mills
smothering, 57, 58
Sorby, Dr H. C., 74
Spencer Steelworks, 98, 100
spray steelmaking, process of, 101

Staffordshire Thick, 24
Staffordshire tuyere, 43
stainless steel, *see* steels
Stanners Closes Steel Company, 121
Stanton Ironworks, 121
steam hammer, *see* hammers
steel castings, 120–1
steel, definition of, 6
steel forges, 134–5
steel forgings, 132–4, 135
steel reinforcement, xi
steels
   blister, xi, 70
   mild, xvi, 3, 7, 40, 49, 69, 71, 80, 107, 132
   carbon, 7, 9, 69, 70, 71, 74, 89, 96, 107, 132
   alloy, 8, 9, 89, 90, 91, 96, 106, 107, 108, 109, 132, 134
   stainless, 9, 90, 91
   shear, 70
   double shear, 70
   crucible, 70, 107
   tungsten alloy, 74, 89
   manganese, 89, 90
   silicon, 90
Stocksbridge Steelworks, 106
Sturtevant, S. *A Treatise of Metallica*, xvi
swarf, 91
Swedish Kaldo process, 101

tap-hole gun, 85
Ten Yard coal, 24
Thomas, Sidney Gilchrist, 78, 79
three-high mill, *see* rolling mills
Tijou, Jean, 127
tilt hammer, *see* hammers
tinplate, 94, 95, 96
titanium, 108
tonnage oxygen, 99
tonnage steels and steelmaking, 71, 74, 79, 87, 101, 109
tungsten alloy steels, *see* steels
two-high mill, *see* rolling mills

upset forging, 128

INDEX

vacuum degassing, 109–11
vacuum melting, 108
vacuum remelting, 108–9
Vickers Works, Sheffield, 136

Walter Somers Ltd, 135
waste-heat recovery, 50, 75
waste-heat steam generation, 50
water power, 14, 19, 20
  shortage of, 29
Watt, James, 30
  and Watt's engine, 30–2
  and Boulton and Watt Soho partnership, 30, 31
  and Boulton and Watt correspondance, 117
welding, 4, 5, 6, 125, 126, 127, 128
wet puddling, see puddling
While, Charles, 81
white iron, see irons
Whitehouse, Cornelius, 125
Whitworth, Joseph, 133

Wilkinson, Isaac, 117
Wilkinson, John, 30–1, 118
Wilkinson, William
  and cupolas, 118
Willenhall Furnaces, 83
wrought iron, xi, xv, xvi, 9, 28, 38, 40, 50, 59, 69, 70, 71, 73, 125, 126, 127, 128, 134
  description of, 3–5
  ease of welding, 4
  corrosion resistance, 5
  and Britannia Bridge, Menai Strait, 9
  and Royal Albert Bridge, river Tamar, 9
  origins of, 10–11
  and dry puddling process, 32–4, 46
  and wet puddling process, 46–9
  fall in output of, 55
  reworking of, 60–1
  products, 61–5